ENSURING EXCELLENCE

ENSURING EXCELLENCE

Mastering LTQR in Nuclear and Defence Manufacturing

*A Practical Guide for
Quality Professionals*

GRAZINA PAVTEL

Contents

Foreword

In today's rapidly evolving industrial landscape, the importance of maintaining a consistent and robust approach to quality management has never been more crucial. As we strive for innovation, efficiency, and safety, the need for meticulous documentation and rigorous standards becomes evident. This is where Life Time Quality Records (LTQR) play a vital role.

This book offers a comprehensive exploration of LTQR, providing valuable insights, best practices, and expert guidance to help professionals effectively create, manage and leverage Lifetime Quality Records. From the importance of traceability and concurrent signing of inspection and test plans to the specific applications in sectors like nuclear and defence, this book delves deep into the practicalities and benefits of maintaining thorough and accurate quality records throughout a product's lifecycle.

What makes this book unique is its balanced approach, combining the foundational principles of LTQR with their real-world application in today's complex industrial environments. It serves as a comprehensive guide for beginners, a trusted reference for experienced quality engineers, and a valuable strategic resource for managers and decision-makers. This book is written in a clear, fun, and engaging style, designed to make learning enjoyable and far from boring. Complex concepts are broken down in an accessible way, ensuring that the content is both informative and captivating.

This book will guide you through everything you need to know to establish, maintain, and leverage LTQR to support your quality management efforts for critical products in highly regulated industries such as nuclear and defence.

This book is a comprehensive guide to mastering Lifetime Quality Records (LTQR) in nuclear and defence industries, with a particular focus on overhead crane manufacturing. However, the principles, insights, and strategies discussed here are universally applicable to any manufactured product within these highly regulated sectors. As you explore the chapters ahead, I encourage you to approach them with curiosity, openness, and a commitment to excellence. The knowledge shared in these pages is designed to equip you with the precision and expertise needed to navigate the complexities of quality management with confidence.

Preface

The concept of Life Time Quality Records (LTQR) may seem straightforward at first glance, but its significance permeates every aspect of quality management in industries worldwide. As a tool, LTQR embodies the meticulous documentation of all quality-related activities throughout a product's entire lifecycle – from design and manufacturing to operation and decommissioning. This book was born out of a deep understanding of how vital these records are, not only for compliance and traceability but also for fostering a culture of continuous improvement, accountability, and excellence.

I was inspired to write this book after noticing a recurring gap in the understanding and application of LTQR practices across various industries. Despite their crucial role in ensuring product safety, regulatory compliance, and customer satisfaction, comprehensive resources on effectively implementing and maintaining these records are often lacking. This book aims to bridge that gap by offering a detailed exploration of LTQR principles, practical guidance, and real-world case studies within the nuclear and defence sectors.

Throughout the chapters, you will find details on the foundational elements of LTQR, including the importance of traceability, the role of quality engineers, and the benefits of maintaining a well-structured quality record system.

The goal of this book is to serve as both a guide and a reference for professionals at all levels – from beginners looking to understand the basics of LTQR to quality managers aiming to optimise their existing practices. The content is designed to be practical and relevant, providing readers with actionable strategies, best practices, and lessons learned from real-world case studies.

Writing this book has been a journey of exploration and learning. I have drawn upon my experiences and the collective wisdom of many industry experts who have generously shared their insights and expertise. I hope this book will be a valuable resource that empowers you to implement and sustain effective LTQR practices within your organisation, driving both quality and operational excellence.

Acknowledgements

Writing this book on Lifetime Quality Records (LTQR) has been an incredibly rewarding journey, and it would not have been possible without the support, guidance, and encouragement of many people and organisations. I would like to take this opportunity to express my deepest gratitude to everyone who has contributed to the development of this book.

First and foremost, I would like to thank my colleagues whose expertise and passion have been a constant source of inspiration. Your insights, experiences, and feedback have shaped the ideas and concepts within these pages, and I am grateful for the knowledge you have shared with me over the years.

A heartfelt thank you to Danny Pickard, the CEO of Kinetic Solutions Group (SCX Special Projects) for sponsoring the publishing of this book. Your support has been instrumental in bringing this project to life, and I am truly grateful for your belief in the value of this work.

To my colleagues in the crane, nuclear, and engineering industries, especially Brian Wilson, Martin Murray, Prof Amanda McKay, Matt Boden, Damien Johnson and Craig Mullen, thank you for sharing your experiences and challenges with LTQR. Your contributions have provided a wealth of practical insights that will help others navigate the complexities of quality management in their own sectors.

A special thanks to my husband, friends and family for their unwavering support and patience throughout the writing process. Your encouragement and understanding gave me the motivation I needed to keep pushing forward, even on the most challenging days.

Lastly, I extend my gratitude to all the readers and quality professionals who are committed to fostering a culture of continuous improvement and excellence. This book is for you, and I hope it serves as a valuable resource in your journey toward achieving the highest standards of quality in your work.

Thank you for embarking on this journey with me.

Format of the book

In this book, you'll find interesting facts and thought-provoking Food for Thought elements. These elements are designed to help you see the bigger picture, connect complex concepts, and deepen your understanding of Lifetime Quality Records (LTQR) and their applications in industrial settings.

Throughout the chapters, you'll encounter the following symbols that highlight key insights:

Represents interesting facts.

Look for these symbols to uncover intriguing, lesser-known details that can spark your curiosity and expand your perspective.

Represents tips for best practices.

Keep an eye out for these tips to optimise your approach.

Represents an insightful fact or idea.

When you see this symbol, it's a moment of clarity – an idea you can directly apply to your workplace.

Represents Food for thought – encourages reflection.

This symbol invites you to pause and reflect on deeper concepts and your personal growth.

Represents a touch of excitement.

This symbol is used to highlight something special or noteworthy, drawing attention to fun and engaging details throughout the book. It adds a spark of excitement to the information, making it more enjoyable and memorable.

By paying attention to these symbols, you'll be able to navigate through the content with a focus on expanding your knowledge and integrating these principles into your professional practice.

CHAPTER 1

IN INDUSTRIES WHERE PRECISION, safety, and reliability are non-negotiable, such as nuclear and defence, the manufacturing process must adhere to the highest standards of quality and accountability. Lifetime Quality Records (LTQR) play an essential role in achieving these standards. This chapter explores the integration of LTQR into the manufacturing processes for nuclear and defence applications, detailing how it drives excellence, facilitates regulatory compliance, and supports continuous improvement. The chapter also discusses the structure of the supply chain, from Tier 1 to Tier 3 suppliers, and how LTQR support risk management, accountability, and continuous improvement. Additionally, it explores key UK nuclear and defence sites, illustrating the role of LTQR in maintaining operational integrity across these critical locations.

LTQR

*Where every detail is a commitment,
and every signature is accountability.*

LTQR – Safeguarding quality and compliance in the Nuclear and Defence sectors

Imagine a world where every bolt, weld, and inspection is meticulously recorded – a world where every piece of critical equipment is traceable from inception to decommissioning. In the high-stakes arenas of nuclear and defence manufacturing, this is not just a vision but a reality, thanks to Life Time Quality Records (LTQR). Manufacturing for the nuclear and defence sectors presents unique challenges. Components and systems must operate flawlessly under extreme conditions while meeting rigorous regulatory requirements. LTQR provides a comprehensive framework to document every phase of production – from design and fabrication to testing and maintenance. This detailed record-keeping not only ensures product integrity but also instils confidence among stakeholders, including regulatory bodies, clients, and end-users.

1.1 STRICT COMPLIANCE REQUIREMENTS

Both nuclear and defence industries are governed by stringent regulations and standards. Compliance with international and national standards is critical for operational safety and legal certification. LTQR plays a pivotal role in:

- **Demonstrating regulatory adherence:** Comprehensive records facilitate audits and inspections, enabling manufacturers to prove that every component meets required safety and quality criteria.

- **Traceability:** Detailed documentation of every material, process, and inspection step ensures that every part can be traced back through the production process, a vital factor during regulatory reviews.

1.2 KEY UK SITES WHERE LTQR IS APPLIED

The United Kingdom is home to several vital nuclear and defence sites, each playing a crucial role in the nation's security, energy infrastructure, and technological advancement. These sites, spread across the country, are not just facilities – they represent the backbone of the UK's defence capabilities and its ambitions in energy and innovation. Let's explore these key locations and their significance, along with the indispensable role of LTQR in supporting them.

Aldermaston and Burghfield (Atomic Weapons Establishment – AWE):

At the heart of the UK's nuclear deterrent sits Aldermaston and Burghfield, where the design and maintenance of nuclear weapons systems are undertaken. These facilities ensure that the nation's nuclear arsenal remains at the highest standard of safety, reliability, and effectiveness. AWE's role in the country's defence strategy is paramount, as it safeguards national security through advanced technology and continuous innovation.

Sellafield:

Situated on the coast of Cumbria, Sellafield is one of the most well-known nuclear sites in the UK. It's primarily focused on fuel reprocessing and decommissioning, playing a vital role in managing nuclear waste and reducing the environmental footprint of nuclear power. With an eye towards the future, Sellafield also contributes to clean energy efforts, marking it as a leader in sustainable nuclear technology.

Rolls-Royce Submarines (Derby):

Known for its iconic engineering prowess, Rolls-Royce's submarine division in Derby is a key player in nuclear propulsion systems for the Royal Navy. They design and build the reactors that power the UK's nuclear submarines – vessels that provide unmatched strategic capabilities and national security. Their work is a testament to cutting-edge technology, precision, and safety.

BAE Systems (Barrow-in-Furness):

In Barrow-in-Furness, BAE Systems specialises in constructing nuclear submarines, which are vital to the UK's defence and deterrent strategy. Their work ensures that the UK's submarine fleet remains robust, reliable, and at the forefront of global military capabilities. The site is integral in maintaining the UK's position as a world leader in naval defence.

Devonport Royal Dockyard:

The historic Devonport Royal Dockyard in Plymouth is home to the UK's naval fleet maintenance operations, including essential submarine refits and repairs. The Dockyard provides critical support to the Royal Navy, ensuring that submarines are always operational and ready for any task. Its strategic role extends far beyond basic maintenance – ensuring that the UK's nuclear-powered vessels remain at peak performance.

Culham Centre for Fusion Energy:

The Culham Centre for Fusion Energy is where the future of nuclear energy is being forged. This site is a global leader in the quest to unlock the potential of nuclear fusion – a clean, virtually limitless energy source that could revolutionise the world's energy supply. By harnessing the power of the stars, Culham is driving forward an ambitious vision for a sustainable energy future.

Together, these sites form an intricate network that plays an essential role in ensuring the UK's leadership in nuclear and defence manufacturing. But these facilities don't operate in isolation. They depend on a robust framework of safety, regulatory compliance, and operational excellence. This is where Life Time Quality Records (LTQR) come into play.

The Importance of LTQR

Life Time Quality Records (LTQR) are critical in these industries because they track the quality, performance, and safety of materials and systems throughout their entire lifecycle. In industries where precision and reliability are essential – such as nuclear and defence manufacturing – LTQR ensure that every action, decision, and change is fully documented and traceable.

These records are far more than just paperwork – they are the foundation of long-term safety and compliance. For instance, at Sellafield, LTQR are vital in managing the reprocessing of nuclear fuel and the secure disposal of waste. Every stage of the process is carefully documented to ensure that all safety and environmental standards are met, and these records are preserved for future reference to guarantee long-term accountability.

At AWE, LTQR support the ongoing development and maintenance of nuclear weapons systems. They ensure that the components of these highly sensitive systems remain up to date and are consistently reviewed to meet national and international safety standards. The longevity and accuracy of these records are paramount in avoiding potential risks and guaranteeing operational integrity.

In a complex environment like Rolls-Royce Submarines or BAE Systems, LTQR track every part of a nuclear submarine's design, construction, and operation. With highly specialised materials and technology involved, these records offer a detailed history of performance, inspections, and maintenance – vital for the ongoing safety and performance of the UK's military assets.

Why Manufacturers Must Compile LTQR

For manufacturers operating in such high-stakes industries, compiling LTQR is not just a regulatory requirement – it is an absolute necessity. The process ensures that the manufacturers maintain a detailed and accurate record of every part, every system, and every procedure over the lifetime of the product. This documentation guarantees that manufacturers can verify compliance with safety standards, provide transparent evidence of product history, and protect against any potential failure or risk.

Moreover, LTQR allow for the continuity of operations even when changes in personnel or technology occur. When systems are upgraded, parts replaced, or procedures altered, these records serve as the reference point that ensures quality and safety are never compromised. In the event of an inspection, audit, or issue arising, LTQR provide manufacturers with the tools to demonstrate that all processes and products have met the highest standards.

In short, the UK's nuclear and defence sites are not only essential for national security and energy innovation – they also rely on LTQR to keep operations running smoothly, safely, and within the scope of strict regulatory standards. These records, preserved over time, ensure that each site's operations are not only traceable but also compliant with the highest standards of quality. They are the key to maintaining long-term safety, minimising risks, and supporting operational efficiency across the country's most critical infrastructures.

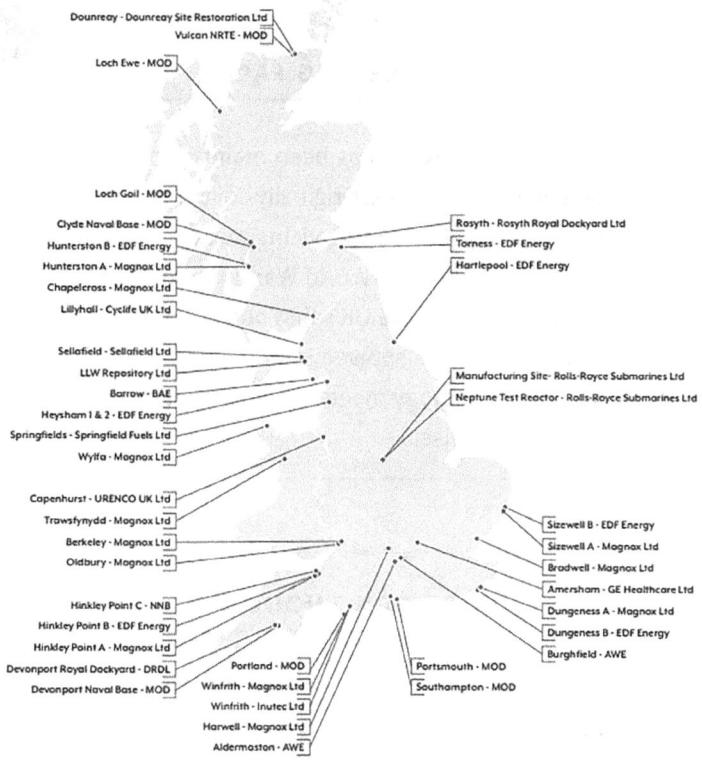

Office for Nuclear Regulation

Map of regulated sites/facilities

Dounreay - Dounreay Site Restoration Ltd
Vulcan NRTE - MOD
Loch Ewe - MOD

Loch Goil - MOD
Clyde Naval Base - MOD
Hunterston B - EDF Energy
Hunterston A - Magnox Ltd
Chapelcross - Magnox Ltd
Lillyhall - Cyclife UK Ltd

Rosyth - Rosyth Royal Dockyard Ltd
Torness - EDF Energy
Hartlepool - EDF Energy

Sellafield - Sellafield Ltd
LLW Repository Ltd
Barrow - BAE
Heysham 1 & 2 - EDF Energy
Springfields - Springfield Fuels Ltd
Wylfa - Magnox Ltd

Manufacturing Site- Rolls-Royce Submarines Ltd
Neptune Test Reactor - Rolls-Royce Submarines Ltd

Capenhurst - URENCO UK Ltd
Trawsfynydd - Magnox Ltd
Berkeley - Magnox Ltd
Oldbury - Magnox Ltd

Sizewell B - EDF Energy
Sizewell A - Magnox Ltd
Bradwell - Magnox Ltd
Amersham - GE Healthcare Ltd
Dungeness A - Magnox Ltd
Dungeness B - EDF Energy
Burghfield - AWE

Hinkley Point C - NNB
Hinkley Point B - EDF Energy
Hinkley Point A - Magnox Ltd
Devonport Royal Dockyard - DRDL
Devonport Naval Base - MOD
Portland - MOD
Winfrith - Magnox Ltd
Winfrith - Inutec Ltd
Harwell - Magnox Ltd
Aldermaston - AWE
Portsmouth - MOD
Southampton - MOD

MOD - Ministry of Defence
DRDL - Devonport Royal Dockyard Ltd
EDF Energy - EDF Energy Nuclear Generation Ltd
AWE - Atomic Weapons Establishment Plc
BAE - BAE SYSTEMS Marine Ltd
NNB - NNB GenCo HPC Ltd

March 2022

www.onr.org.uk

By prioritising LTQR, the UK is better equipped to meet future challenges, secure its defence interests, and advance its global position in nuclear technology. The map shows site locations as at March 2022.

INTERESTING FACT

Devonport Royal Dockyard has been maintaining the Royal Navy's fleet for over 300 years. Originally built in the late 1600s, it now plays a crucial role in servicing nuclear submarines. Similarly, Sellafield, once a World War II munitions factory, later became the site of the UK's first nuclear power station. These historic sites have evolved into cutting-edge facilities, where modern technology meets decades – even centuries – of engineering expertise.

FOOD FOR THOUGHT

I hope this book not only provides you with a deeper understanding of LTQR but also inspires you to consider a career within these industries. Working in the nuclear and defence sectors offers a unique opportunity to contribute to projects of national and global importance, where precision, safety, and innovation are paramount. Whether your expertise lies in engineering, quality assurance, regulatory compliance, or operations, there is a vital role for dedicated professionals like you in these world-class facilities.

1.3 TIERS IN THE NUCLEAR AND DEFENCE SUPPLY CHAIN

To truly grasp how manufacturing works in the UK's nuclear and defence sectors, it's crucial to understand the different tiers within the supply chain. Each tier plays a vital role in ensuring that every part of the system meets the highest standards of quality, safety, and compliance. Let's break down the tiers and explore how they work together to create some of the most advanced and reliable systems in the world.

Tier 1 Suppliers: The Backbone of the system

Tier 1 suppliers are the powerhouses of the nuclear and defence supply chain. These are the primary contractors responsible for delivering fully integrated systems and critical components, often working directly with governments, regulatory bodies, and major industry players. They oversee large-scale projects and ensure that every deliverable meets stringent safety and quality standards. Their work is foundational – without them, the entire system would lack the framework to operate safely and effectively. These suppliers deal with the biggest projects and ensure the complex systems that power submarines, nuclear reactors, and military technology come together seamlessly.

Tier 2 Suppliers: Specialising in precision

Moving one step further down the chain, we find the Tier 2 suppliers. These companies provide the specialised components, subsystems, and services that help complete larger assemblies. They may not be handling the entire system like Tier 1, but their contributions are no less critical. Think of them as the experts in the details – the ones who supply the parts that integrate directly into Tier 1 systems. Whether it's a highly specialised valve for a nuclear reactor or the complex electronics that power military systems, Tier 2 suppliers' products must meet the same high standards. For these companies, adherence to LTQR is essential to ensure traceability, consistency, and quality across the entire chain.

Tier 3 and Below: The Foundations of Manufacturing

At the foundation of the supply chain, we have Tier 3 suppliers and below, responsible for providing the raw materials, fasteners, electronic components, and other fundamental elements that go into the final products. While these suppliers may seem more removed from the big-ticket systems, their role is just as vital. The nuts and bolts, wires and metals, and every seemingly small part must be of the highest quality because even the smallest flaw could have a significant impact on the final product. This is why Tier 3 suppliers, though further down the chain, must also maintain rigorous LTQR practices. By doing so, they ensure that quality and traceability flow seamlessly from one tier to the next, right up to the final product.

"There has never been a greater need for Lifetime Quality Records – essential for traceability, compliance, and safety – to drive efficiency and excellence."

The importance of quality flow across the Tiers

The relationship between these Tiers is more than just a supply of parts – it's a carefully coordinated flow of quality assurance and documentation. Each level must uphold the same stringent standards to maintain the integrity of the entire supply chain. If there's a deviation or lapse in LTQR at one Tier, it can create a ripple effect throughout the supply chain, compromising the reliability and safety of the final product. For example, a small issue with a component from Tier 3 could potentially cause delays or safety risks in Tier 1's final systems. That's why every supplier, no matter their Tier, must be fully committed to quality, traceability, and the documentation that supports it.

In essence, the nuclear and defence supply chain is built on trust and meticulous attention to detail at every level. When Tier 1, Tier 2, and Tier 3 suppliers work in unison, maintaining robust LTQR systems and adhering to the highest standards, the result is a powerful, reliable, and safe system that powers the UK's nuclear and defence capabilities. The integrity of each part of the supply chain directly impacts the success of the entire ecosystem, and it's only through seamless collaboration and strict quality assurance that these industries can continue to thrive.

INTERESTING FACT

A single defective component from a Tier 3 supplier can have massive consequences across the entire nuclear and defence supply chain. Even the smallest flaw – like an impurity in a bolt or a miscalibrated sensor – can lead to costly delays, safety risks, and system failures at the highest levels. That's why every Tier, from raw material providers to top-tier integrators, must maintain rigorous LTQR practices. Quality isn't just a Tier 1 responsibility – it's a shared commitment that ensures the integrity of the entire system.

1.4 RISK MANAGEMENT AND ACCOUNTABILITY

In the high-stakes environments of nuclear and defence manufacturing, even minor oversights can have severe consequences. LTQR serves as an indispensable tool for:

- **Risk Identification:** Systematic documentation helps identify potential issues early in the production process.

- **Mitigation Strategies:** With detailed records, Quality Engineers can implement timely corrective actions, reducing the risk of costly recalls or operational failures.

- **Accountability:** Every entry in an LTQR is backed by the expertise of qualified professionals, whose signatures transform data into a personal commitment to excellence.

1.5 BUILDING A STRONG QUALITY FRAMEWORK

LTQR transforms the manufacturing process into a seamless, efficient system, ensuring every step is precise and purpose-driven. It begins right at the drawing board, where careful planning and attention to detail lay the foundation for quality at every stage.

Design documentation: Every design decision, material choice, and engineering calculation is recorded. This ensures that even before production begins, there's a clear blueprint of what the final product should be.

Controlled processes: As manufacturing kicks off, LTQR captures the entire process – from precision machining and high-stakes welding to the final assembly and finishing touches.

Rigorous testing: Before any component can be deployed, it undergoes thorough testing. LTQR logs every inspection and test result, confirming that the product not only meets but exceeds all operational requirements.

1.6 CONTINUOUS IMPROVEMENT THROUGH LTQR

LTQR is a living document that grows with each project. By analysing the data collected, manufacturers can:

- **Identify process bottlenecks:** Insightful data analysis reveals inefficiencies and areas for improvement.

- **Implement best practices:** Lessons learned from previous projects are incorporated into new designs and production processes, fostering a culture of continuous improvement.

- **Enhance product reliability:** Continuous refinement based on LTQR feedback leads to more reliable, higher-performing products.

1.7 THE ROLE OF QUALITY ENGINEERS IN LTQR IMPLEMENTATION

Quality Engineers are at the forefront of LTQR management. Their responsibilities include:

- **Regulatory interpretation:** Ensuring that every aspect of the LTQR meets evolving nuclear and defence standards.

- **Risk assessment:** Identifying potential risks through detailed analysis of the records and taking proactive measures.

- **Validation and sign-off:** Quality Engineers add their signatures to LTQR documents, symbolising personal accountability and ensuring that each record reflects a high standard of quality and safety.

1.8 ACCOUNTABILITY AND SIGNATURES IN LTQR

A critical element of LTQR is the Quality Engineer's signature. This signature is more than a formality – it represents:

- **Personal commitment:** It signifies that the engineer has thoroughly reviewed the documentation and stands by its accuracy.

- **Authority and competence:** Only those with the requisite expertise and authority sign off on the LTQR, thereby ensuring that all documented processes have met strict quality standards.

- **Trust and credibility:** A signed LTQR enhances transparency and instils confidence among stakeholders, reinforcing the manufacturer's commitment to excellence.

1.9 THE PEOPLE, PROCESS, AND TECHNOLOGY

LTQR is more than just a system – it's the product of a powerful synergy between people, process, and technology. Industries as diverse as aerospace, defence, nuclear, energy, automotive, pharmaceuticals, rail, and construction rely on LTQR principles to maintain excellence. It's this dynamic interplay that transforms LTQR from a set of records into a vibrant engine of quality and safety, driving continuous improvement across high-stakes industries.

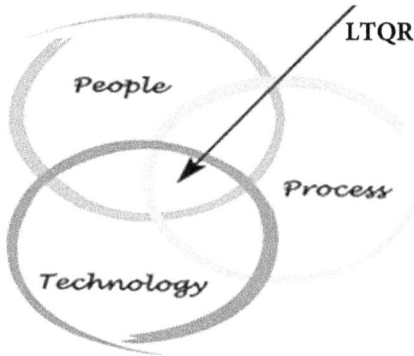

People

Process

Technology

LTQR

INTERESTING FACT

Companies that rigorously implement LTQR practices in their manufacturing processes have observed a significant reduction in unplanned downtime and maintenance incidents – sometimes as much as 30%. This robust documentation not only ensures compliance with stringent regulatory requirements but also acts as an early-warning system, enabling proactive identification and mitigation of potential issues, thereby safeguarding both operational excellence and safety.

CHAPTER 2

THIS CHAPTER INTRODUCES THE core concepts of Lifetime Quality Records (LTQR) within the crane industry. It explains what LTQR are and why they are vital for crane manufacturers, operators, and service providers. Far from being just paperwork, LTQR are detailed, dynamic records that track products' entire life cycle – from design and production to maintenance and operation.

Key principles are outlined, emphasising how LTQR ensure safety, regulatory compliance, and operational efficiency. The chapter highlights the essential components of an LTQR, such as design specifications, manufacturing records, testing data, and maintenance records. By maintaining comprehensive and accurate LTQR, companies can guarantee that their products and services consistently meet the highest industry standards, ensuring long-term reliability and performance.

"

You can't manage what you don't measure, and you can't measure what you don't document.

CHAPTER 2

Introduction to LTQR

Having worked in quality roles with manufacture of overhead cranes, I've seen first-hand how the principles of quality and traceability that support crane manufacturing extend well beyond just cranes. While this book focuses on the specific application of Lifetime Quality Records (LTQR) in overhead crane manufacturing – particularly in nuclear and defence industries – the framework we explore here is not exclusive to cranes. In fact, the same LTQR principles apply seamlessly to a wide range of critical components across industries. Whether we're talking about reactors, pipes, vessels, electrical components, or other systems, the approach to maintaining rigorous quality records and ensuring full traceability remains consistent. Just as LTQR are vital to the safe and reliable operation of overhead cranes, they are equally essential for ensuring the integrity and performance of other high-stakes equipment.

By the time you finish reading this book, you'll be equipped to apply the same LTQR framework to your own work, regardless of the industry or the component you're working with.

Overhead cranes in nuclear and defence applications serve critical roles, whether they're lifting reactor components, handling hazardous materials, or transporting sensitive defence equipment. Even the smallest defect or oversight in their design or operation can lead to catastrophic safety risks, financial losses, or operational downtime. LTQR mitigate these risks by providing a comprehensive, documented trail of quality data that spans every phase of a crane's lifecycle.

From missile transportation cranes to sensitive equipment handlers, LTQR serve as a comprehensive roadmap, tracing the origin, manufacture, and operational history of every component. This level of traceability is indispensable in an industry where failure is not an option and where global security often hangs in the balance.

Picture this: You're in a busy factory, and an overhead crane is lifting a massive 50-ton steel beam. It moves with incredible precision, swinging smoothly and safely across the room. It looks effortless – but behind this moment lies a mountain of work, countless safety checks, and an invisible but powerful ally – Lifetime Quality Records (LTQR).

LTQR are like a crane's life story, recording everything from the raw materials it's made from to the last inspection it went through. They're not just boring paperwork – they're the key to ensuring that every crane performs safely and reliably, day in and day out. Think of them as the "black box" of crane manufacturing: if something goes wrong, LTQR hold the answers.

Let's dive in and see how these records keep overhead cranes in top shape, protect workers, and make sure manufacturers sleep soundly at night.

> ### IDEA
>
> Imagine that same level of detail and traceability applied to the components your company works with, whether it's reactors, pipes, or other critical systems.

2.1 WHAT ARE LIFETIME QUALITY RECORDS (LTQR)?

Lifetime Quality Records (LTQR*) are detailed and comprehensive records that capture every aspect of product quality and safety from its initial design and engineering phases through its active operational life, including maintenance and inspections, all the way to its retirement or decommissioning. These records serve as a complete historical account of all data, processes, activities, inspections, and modifications that have a direct or indirect impact on product quality and performance.

Think of the LTQR as a "digital passport" for each product/ component – a dynamic, living record that provides a clear and detailed history of every aspect related to quality assurance and control, helping to ensure that the product meets all required standards, both now and in the future.

---

* Also known as LTR and LQR. In this book, I will refer to it as LTQR, as this is the term most commonly used within the nuclear and defence sectors.

Key components of LTQR

- Project Documentation

- Design Documentation

- Manufacturing Documentation

- Proprietary Equipment Documentation

- Assembly & Testing Documentation

2.2 SCOPE OF LTQR IN OVERHEAD CRANE MANUFACTURING

To fully understand the scope of LTQR, it's important to break down the different types of records that contribute to a complete LTQR. Each of these records is crucial for ensuring the safe, reliable, and efficient operation of cranes throughout their entire lifecycle.

Imagine LTQR as a crane's personal diary, documenting every phase of its life, from its initial design to its eventual decommissioning. The scope of LTQR in crane manufacturing covers every stage of the crane's journey. It provides a 360-degree view, ensuring that at every step, the crane meets the highest safety and quality standards.

Here's a breakdown of the major areas that LTQR cover in overhead crane manufacturing:

1. Design and Engineering – *The Blueprint stage*

Every great crane starts with a great design, and this is where the LTQR journey begins. The design and engineering phase of a crane is critical because it lays the foundation for everything that follows. An LTQR captures every design detail in this stage:

Manufacturing drawings
Detailed drawings of the crane's structure, components, and layout. These designs are critical for ensuring the crane's ability to perform its job safely and efficiently.

Calculations and specifications
Load calculations, safety margins, material strength, and environmental considerations. These technical documents show that the crane was designed to handle its intended workload and more!

2. Procurement, Manufacturing, and Assembly – *The Construction stage*

Once the design is finalised, the procurement and manufacturing process begins, and this is where things get interesting! LTQR follow every step of crane production, ensuring that every part meets exacting standards. Here's what's covered:

Procurement

Sourcing materials – The procurement process involves selecting and acquiring high-quality materials that meet strict specifications. LTQR document every supplier and the certifications of the materials, such as steel grades, welding rods, and electrical components, ensuring compliance with industry standards.

Vendor selection and approval – Every vendor and subcontractor involved in the crane's production is vetted to ensure they meet quality standards. The LTQR records their qualifications, certification status, and the materials provided.

Steel certification in the crane industry: A quick dive into strength and safety

Steel is the unsung hero behind every crane's ability to lift massive loads with precision and safety. But how do we know the steel in these giants is up to the task? Steel certification is the key.

Why does steel certification matter?

Strength & safety: Certified steel guarantees cranes' crucial parts can handle heavy loads without risk of failure.

Legal compliance: Certification ensures cranes meet all regulatory standards, keeping operations safe and legal.

Top performance: Certified steel ensures the crane remains efficient and durable, keeping it working smoothly for years.

Types of steel certification

Steel comes with different certifications to verify its quality:

- Mill Test Reports: The steel's "birth certificate", proving it meets the required standards.

- Standard Specifications: International guidelines like ISO set the performance bar for steel.

- Third-Party Certification: Independent testing to add extra verification.

- Welding Certification: Ensures that welds holding crane parts together are strong and secure.

Managing steel certification

Maintaining steel certification means staying organised. Keep records up-to-date, foster good supplier relationships, and regularly audit for compliance. It's all about ensuring the steel in use is always top-notch.

Manufacturing Records

Material records: Every material used in the crane, from steel beams to bolts, is tracked. The LTQR includes certificates proving that the materials meet quality standards and are sourced from approved suppliers.

Fabrication logs: The assembly process is documented, including welding, painting, and fitting. This ensures that every component is built correctly and that welding procedures meet industry standards.

Component inspections: As each part of the crane is built, it's inspected and tested for quality. These inspection records are stored in the LTQR to ensure nothing is missed.

Understanding Inspection and Test Plans (ITPs)

Inspection and Test Plans (ITPs) might sound like dry, technical documents, but they're actually the secret sauce to ensuring cranes are built to last! Think of them as a recipe book, carefully outlining the steps, ingredients (tests), and the final checks to ensure your crane isn't just a pile of steel – it's a high-performance, safety-tested marvel ready for action.

What Are ITPs?

An ITP is a detailed guide that specifies which inspections, tests, and verifications are required during the creation of a crane. It tells you who's responsible, when the tests happen, and what the outcome should be. These plans are there to ensure that the crane isn't just built; it's built *right*.

The following table outlines key components of ITPs (think of it as the ultimate crane recipe).

Component	What's happening here?
Inspection points	These are the checkpoints during manufacture and assembly. At each point quality is evaluated like making sure your dough is rising properly before it goes in the oven.
Acceptance criteria	The standards or goals must be hit for the crane to pass the test. It's like the perfect golden-brown crust you need to bake a pie.
Frequency of inspection	Tell you when inspections should happen – whether it's after a certain number of parts are manufactured and assembled or after each test. Like checking on your pie every ten minutes to ensure it doesn't burn.
Responsibilities	These are designated experts who need to conduct the tests. Think of them like the top chefs ensuring your meal (in this case crane) passes all the tests.
Documentation requirements	Everything must be written down: which tests were done, when, by whom, and the outcome. This helps if the crane ever needs a "recipe check" later on.

Why concurrent signing of ITPs is essential for LTQR

Let's talk about why signing ITPs at the moment of testing isn't just a formality – it's a game changer.

1. Real-time verification and accountability

When every inspection or test is signed off right after it happens, you get an immediate record confirming that everything went smoothly – or if there was an issue. It's like checking the temperature of your oven while the pie is baking – ensuring everything's going according to plan!

2. Enhancing traceability and transparency

Every signature is like a stamp of approval, confirming that the inspection happened on time and by the right people. This makes it easy to track back if something goes wrong later – just like going through your recipe book to see if you missed an ingredient.

3. Facilitating timely issue resolution

Immediate signing means immediate feedback! If something's wrong with the crane, you can spot the issue and fix it before it becomes a major problem. It's like realising your pie is starting to burn before it's too late.

A TOUCH OF EXCITEMENT

You've probably noticed that I often compare manufacturing to cooking. This is because I take food seriously, and just as precise ingredients and steps are crucial in cooking, the same level of attention to detail is essential in the manufacturing process. Every element plays a vital role in ensuring the desired outcome, whether in the kitchen or the factory.

4. Supporting compliance with regulations

Many industries have strict rules, and signing off on ITPs ensures you meet all the legal requirements. It's like making sure you follow food safety standards to prevent any health code violations.

5. Improving Quality Assurance

This method keeps quality top-of-mind! With every inspection signed off, everyone knows their role in maintaining the crane's integrity. It's like making sure every step of your recipe is followed precisely for a perfect end result.

Risks of delayed or incomplete signing of ITPs:

Waiting too long to sign off on an ITP or skipping the sign-offs altogether is a recipe for disaster. Here's what can go wrong:

1. Gaps in documentation

Skipping sign-offs can create holes in your records, making it tough to understand exactly what happened – and when. It's like forgetting to write down a step in your recipe and not knowing why your pie didn't turn out quite right.

2. Reduced accountability

Without real-time signatures, it's harder to pin down who was responsible for each test. This could lead to confusion and even blame-shifting when things go wrong. It's like not knowing who added too much salt to your dough.

3. Increased risk of non-compliance

If you don't keep proper records, you could end up missing important legal requirements, leading to fines or trouble down the line. This is like skipping a critical ingredient in your recipe and then realising halfway through that the pie isn't going to taste quite right.

4. Delays in issue resolution

If issues aren't spotted right away, they'll snowball and cause even bigger problems. It's like not catching that your pie is overcooked until it's already too late.

Best practices for concurrent signing of ITPs:

Now that we know why concurrent signing is essential, here's how to do it like a pro:

1. Establish clear procedures and protocols

Make sure everyone knows how and when they should sign off on ITPs. This is like setting clear cooking instructions so everyone in the kitchen is on the same page.

2. Use digital tools and technology

Leverage modern technology, like digital signatures, to speed up and secure the process. It's like using a smart oven that lets you track your pie's temperature from your phone.

3. Train and empower personnel

Get everyone involved in quality checks trained up and motivated to sign off as they go. Think of it like teaching your team how to bake the perfect pie every time.

4. Conduct regular audits and reviews

Regularly check if your ITPs are being signed properly and on time. This is like doing a taste test throughout the baking process – always checking if something needs tweaking.

5. Foster a culture of Quality and Accountability

Encourage a culture where everyone takes pride in their work and understands the importance of signing off at the right time. This ensures that every "slice" of the crane comes out perfectly, just like a top-notch pie.

Dimensional Records

In the crane industry, precision is everything. Whether lifting a few tons or several hundred, every component, beam, and bolt must meet exacting standards to ensure safety, efficiency, and longevity. One of the fundamental aspects of maintaining these standards is the meticulous documentation of Dimensional Records.

What are Dimensional Records?

Dimensional records are detailed documents that capture the precise measurements, tolerances, and physical attributes of all crane components, from the smallest bolt to the largest structural beam. These records include dimensions such as length, width, height, diameter, thickness, and the distance between holes, among others. Essentially, they are the blueprints for ensuring that every part of the crane is manufactured, assembled, and maintained according to exact specifications.

This table provides a clear overview of Dimensional Records, their significance, and how they are used within the crane industry to maintain safety, efficiency, and quality throughout the crane's lifecycle.

Category	Details
Importance	*Safety compliance.* Ensures proper fitting and alignment, reducing failure risks.
	Quality assurance. Provides a benchmark for inspections and testing.
	Maintenance and repairs. Helps with sourcing and fabricating replacement parts.
	Lifecycle management. Tracks changes and modifications.
Components	*Part identification.* Unique identifier for each component.
	Measurement data. Accurate dimensions (length, width, height).
	Tolerances. Permissible variation limits for dimensions.
	Material specifications. Information about material grade, treatment, and finish.
	Inspection records. Data from measuring tools and inspections.
Benefits	Helps maintain crane safety, quality and longevity, reduces downtime, lowers repairs costs and ensures regulatory requirements.

Calibration records

Table below summarising equipment calibration and LTQR. This table captures the essence of equipment calibration and its integration with LTQR, emphasising its role in ensuring accurate, reliable, and compliant product manufacturing and maintenance.

Category	Details
Introduction	Equipment calibration ensures measurement accuracy, reliability, and consistency, and is key to maintaining quality over a product's lifecycle. It ties into LTQR to ensure quality and regulatory compliance at each step of the manufacturing process.
What is equipment calibration?	Calibration is the process of adjusting instruments to measure accurately by comparing them to a known standard, ensuring that measurements are precise and meet global standards.
Why calibration matters?	■ **Accuracy:** Ensures correct measurements, vital for product quality and safety. ■ **Consistency:** Maintains uniformity across production, ensuring identical quality. ■ **Compliance:** Meets regulatory standards, easing audits and inspections. ■ **Reliability:** Reduces defects and failures, ensuring consistent product performance.

Calibration & LTQR integration	Calibration checks become integral parts of LTQR, documenting the accuracy and reliability of equipment used throughout the product's lifecycle. Each calibration adds an entry to LTQR, supporting product quality and compliance.
What should calibration records include?	■ **Documentation:** Records should detail the date, results, standards, and any adjustments made. ■ **Retention:** Records must be stored for an extended period, often for the equipment's entire lifecycle. ■ **Accessibility:** Records should be easy to access for audits, inspections, and reviews.
Integrating calibration into workflow	■ **Scheduled Calibration:** Calibration should be routine, like a regular check-up, integrated into LTQR workflows. ■ **Automatic Updates:** Modern LTQR systems can automate recalibration schedules and notify relevant personnel to ensure timely calibration.

Consumable Certification Records

Here's a table summarising the consumable certification. This table breaks down the essentials of consumable certification, highlighting its importance, the types of consumables involved, and the key elements that contribute to quality assurance and compliance.

Category	Details
Definition	The process of verifying and documenting that consumables (welding rods, lubricants, bolts, paints, gases) meet specific standards and requirements.
Importance	Ensures safety, reliability, and compliance with regulations. Helps maintain crane performance and integrity.
Why It Matters?	■ **Safety & reliability:** Prevents failures and accidents. ■ **Regulatory compliance:** Ensures adherence to industry standards. ■ **Quality assurance:** Guarantees consumable quality. ■ **Traceability & accountability:** Tracks materials for issues. ■ **Lifecycle management:** Part of LTQR, helping with repairs and audits.

Types of Consumables	■ **Welding consumables:** Rods, wires, and electrodes. ■ **Lubricants & fluids:** Oils, greases, hydraulic fluids. ■ **Fasteners & bolts:** Bolts, screws, and nuts. ■ **Paints & coatings:** Primers, protective coatings. ■ **Gases & chemicals:** Gases for welding and other chemical materials.
Key Elements	■ **Material identification:** Type, grade, and specifications. ■ **Supplier details:** Supplier information and certifications. ■ **Batch & Lot numbers:** Traceability for defect identification. ■ **Test results:** Proof of compliance with standards. ■ **Expiry dates & Storage conditions:** Details on shelf life and handling. ■ **Compliance with standards:** Confirmation of adherence to industry regulations (e.g., ISO, AWS).
Role in LTQR	Consumable certifications are integrated into the Lifetime Quality Records ensuring a complete documentation trail for safety, maintenance, and performance.

Surface protection certification

Here's a table summarising Surface Protection Certification: This table provides an overview of Surface Protection Certification, highlighting its importance and how it supports the integrity and safety of the equipment/product throughout its lifecycle.

Category	Details
What is surface protection certification?	Formal documentation verifying the quality, types, and application of coatings used to protect product from corrosion and environmental degradation. Certification ensures compliance with industry standards and manufacturer specifications, providing lasting protection.
Importance of certification	■ **Corrosion prevention:** Ensures coatings resist saltwater, moisture, and chemicals. ■ **Safety assurance:** Maintains structural integrity by preventing rust and corrosion. ■ **Compliance:** Meets industry standards (e.g., ISO, NACE) for regulatory approval. ■ **Extended service life:** Minimises wear, reducing repairs and downtime. ■ **Quality assurance:** Verifies proper application methods and materials.
Types of Surface Protection	■ **Paint coatings:** Epoxy, polyurethane, zinc-rich coatings. ■ **Galvanising:** Zinc coating for corrosion resistance.

	■ **Powder coating:** Durable, heat-cured powder for added protection. ■ **Metallic coatings:** Aluminium/zinc coatings for barrier and sacrificial protection. ■ **Specialised coatings:** Anti-fouling, fire-resistant, and chemical-resistant treatments.
Key Certification Elements	■ **Material specifications:** Coating type, batch numbers, and material data. ■ **Surface preparation:** Cleaning, abrasive blasting, or grinding for coating adhesion. ■ **Coating application:** Method (spraying, dipping), thickness, curing process. ■ **Adhesion testing:** Ensures proper bond to surfaces. ■ **Compliance:** Meets industry standards (ISO 12944, NACE).
Best practices	■ **Clear specifications** ■ **Standardised documentation**: use templates for consistency ■ **Regular Inspections:** Ensure coatings meets standards through periodic checks. ■ **Supplier management:** work with certified vendors for quality materials.
Integration with LTQR	Combined with other records, it forms a comprehensive trail that supports traceability, accountability and regulatory compliance ensuring products' longevity and safety.

BEST PRACTICES FOR DOCUMENTATION

Standardised templates. Use forms or digital templates to capture data consistently.

Digital tools. Use software for storage, updates, and sharing.

Regular updates. Keep records updated with modifications and inspections.

Quality control. Cross-check measurements, conduct audits, and review critical components.

Archival and accessibility. Ensure easy access to records for authorised personnel.

Testing and Certification: Proving the Crane's Worth

Once the crane is assembled, it's time for the big test. The testing phase is where the crane proves its mettle and demonstrates that it can handle the heavy lifting it was designed for. LTQR document all these tests:

Load Tests

The crane is tested by lifting its maximum rated load (or even more) to make sure it's up to the job. The LTQR records the results of these tests, including any adjustments made if the crane doesn't pass.

Stress and safety testing

The crane is subjected to extreme conditions to see how it holds up under pressure. Safety features, like emergency brakes and overload protections, are tested, and the results are recorded.

Certifications

Once the crane passes all tests, it's certified by regulatory bodies. These certificates, along with the test data, are added to the LTQR, confirming that the crane is ready for operation.

2.3 FUNDAMENTAL PRINCIPLES OF LTQR

Creating an effective Lifetime Quality Record (LTQR) system is like building the foundation of a skyscraper – it sets the entire structure for the future, ensuring that everything stands strong, secure, and reliable. Just like engineers follow certain principles to design a crane, LTQR follow fundamental rules to ensure a system that keeps cranes operating safely, efficiently, and in compliance. But what exactly are those principles, and how do they impact crane manufacturing and maintenance? Let's break them down in a detailed way, as we explore the key elements that make up a rock-solid LTQR system.

Before we start detailed principles let's take a quick look at the key phases.

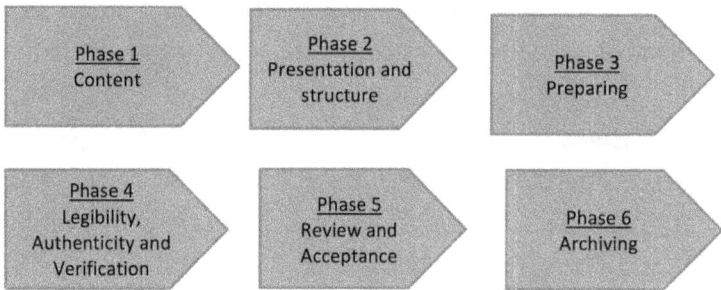

Phase 1 Content	Phase 2 Presentation and structure	Phase 3 Preparing
Phase 4 Legibility, Authenticity and Verification	Phase 5 Review and Acceptance	Phase 6 Archiving

Phase 1 – Content
The content of the LTQR can vary considerably depending upon the nature of the product and customer requirements. Generally during the design phase of the product, the design team identify the necessary records that are required to verify the product has been manufactured to specific requirements.

Requirements for LTQR from the manufacturer are detailed and communicated in the Purchase Order (PO) and associated documentation, e.g. manufacturing specification, quality control plan/ITP.

Phase 2 – Presentation and Structure

LTQR are presented and structured in a consistent manner in accordance with the supplied project LTQR Index.

Generally LTQR presented in A4 lever arch files and labelled with the following information on the front and where possible the spine of the file:

Project Number, Contract Number, Document Reference Number, Equipment/product Title, File No 'x' of 'y', e.g. '1 of 3'.

All sections and sub-sections within LTQR separated using section dividers which are suitably ordered either numerically or alphabetically and correspond directly to the main index.

Phase 3 – Preparing

LTQR prepared concurrent with the manufacturing activities being undertaken and at any point in time provide an up to date accurate record of the current manufacturing status.

LTQR no matter where they are produced in the supply chain are kept as their original package. These are not to be broken into other LTQR packages, they are either included as addendums or given their own unique reference and signposted out to within the main LTQR.

Phase 4 – Legibility, Authenticity and Verification
Documentation contained within the LTQR must be clearly legible, suitable for copying and/or scanning purposes. Original documentation are provided. Where this is not possible, copies of the documentation are taken and endorsed as 'verified true copies of the original'. The use of correction fluid or correction tape to amend any LTQR document is explicitly prohibited at any time. The acceptable method of amendment is a single line through the text in question (leaving that text visible), the correct information entered adjacent to the strikethrough, and the amendment initialled and dated by the person making the change.

Phase 5 – Review and Acceptance
LTQR are made available for review at any time during a visit to a manufacturer's. The completed LTQ are available for review and acceptance prior to the release of any equipment/product. In cases when equipment is subject to a 'staged' release the LTQR are reviewed to ensure they reflect the current status prior to authorising the release of the equipment/product.

During manufacture as the records are being compiled they are maintained and stored in a clean and dry environment, ideally in the office. LTQR should not be held in a production area to avoid the possibility that they could be damaged, destroyed or mislaid. Care should be taken to ensure that the key information is not destroyed by hole-punching. Where drawings are included, they shall be at minimum A3 size for legibility, and folded to A4 size to expose the drawing number.

Unless otherwise stated, following final acceptance the supplier/ manufacturer shall provide the original hard copy plus an electronic (scanned) copy of the whole LTQR package.

Phase 6 – Archiving
LTQR are archived in accordance with the company requirements.

To create an effective LTQR system, it is important to understand and adhere to several fundamental principles:

2.3.1 LTQR Index

The LTQR Index isn't just about ticking boxes – it's about creating a streamlined, organised system that ensures every aspect of a project meets the highest standards. It is structured into nine key sections, each playing a crucial role in documenting quality assurance and compliance. These sections are further broken down to provide detailed guidance for effective project management. The main index sections include:

1. General

This foundational section encompasses essential project requirements, administrative records, and compliance documents, ensuring that all critical information is meticulously documented and readily accessible.

2. Design

Serving as the project's blueprint, this section captures engineering and architectural design elements, including calculations, specifications, and drawings, ensuring alignment with regulatory and project-specific standards.

3. Mechanical (Fabrications and Machining)

Focusing on the mechanical components, this section details design specifications, material selections, fabrication processes, and machining procedures, ensuring all mechanical elements meet safety standards and operational requirements.

4. EC&I

This section addresses electrical, control systems, and instrumentation aspects, particularly control panels, ensuring adherence to safety regulations, performance standards, and operational reliability.

5. Coatings

Beyond aesthetics, this section deals with protective coatings, corrosion prevention, and finishing processes, ensuring applications meet durability and environmental standards.

6. Proprietary Equipment

This section documents equipment sourced from external suppliers, ensuring that all proprietary items meet specified quality and performance criteria.

7. Works Assembly & Testing

Detailing the assembly processes and in-house testing procedures, this section ensures that components are correctly assembled and function as intended before site deployment.

8. Site Assembly & Testing

Focusing on on-site assembly and testing activities, this section verifies that systems and components are correctly installed and operational within the project's environment.

9. Certification

This final section compiles all necessary certifications, including compliance certificates and regulatory approvals, ensuring that the project meets all legal and quality standards.

Each of these sections is further subdivided to cater to specific project needs, ensuring a thorough, organised, and efficient approach to documentation. The LTQR Index isn't just about managing paperwork – it's about building a system that drives quality, compliance, and project success from start to finish.

SECTION 1

Section No.	Contents	Applicable Y or X
1.0	**Project Documentation**	
1.1	Instruction to carry out work e.g. Purchase Order	
1.2	Client Specification	
1.3	Quality Plan	
1.4	Technical Queries / Concessions	
1.5	Non Conformances/ Defect Notices	
1.6	Project Programme	
1.7	Drawing Register (Including As Built Drawings)	
1.8	Manufacturing Specification	
1.9	Coating Procedures	
1.10	Operation & Maintenance Manuals	
1.11	Sub-contractor approvals	
1.12	Miscellaneous Documents	

SECTION 2

Section No.	Contents	Applicable Y or X
2.0	**Design**	
2.1	Inspection & Test Plan (ITP)	
2.2	Basis Of Design (BOD)	
2.3	Cable Block Diagram (CBD)	
2.4	Cable Calculations	
2.5	Cable Schedules	
2.6	Commissioning Spares List	
2.7	Compliance Matrix	
2.8	Control Cable Test Schedules	
2.9	Design Risk Assessment	
2.10	Functional Design Specification (FDS)	
2.11	Design Reviews/changes	

SECTION 3

Section No.	Contents	Applicable Y or X
3.0	**Manufacturing (Fabrications & Machining)**	
3.1	Inspection & Test Plan (ITP)	
3.2	Weld Procedure Specification (WPS) Including Weld Procedure Qualification Record (WPQR)	
3.3	Welder Qualifications	
3.4	NDT/NDE Procedures	
3.5	NDT/NDE Qualifications	
3.6	Material Certification	
3.7	Welding Consumables Certification	
3.8	Welding Traceability i.e. weld maps and weld traceability records	
3.9	Dimensional Inspection Records	
3.10	Calibration Records – Inspection Equipment	
3.11	Declaration of Performance (CE Marking, Execution class)	
3.12	Certificate of Conformity (for all parts, e.g. gearboxes, drive shafts, couplings, actuators – hydraulic, pneumatic, mechanical, pumps, vessels, etc)	
3.13	Inspection Release Certificate (IRC)	
3.14	Delivery Note(s)	

3.15	Load/Pressure Testing	
3.16	Fasteners certification	
3.17	Heat treatment records	

SECTION 4

Section No.	Contents	Applicable Y or X
4.0	**EC&I Manufacturing – (e.g. Control Panels, transformers, cubicles)**	
4.1	Inspection & Test Plan (ITP)	
4.2	Component Data Sheets	
4.3	Component Sub-Orders Purchase Order (un-priced)	
4.4	Component Sub-Orders Certificate of Conformity	
4.5	Test Records	
4.6	Inspection Records	
4.7	Calibration Records – Test Equipment	
4.8	Certificate of Conformity	
4.9	Inspection Release Certificate (IRC)	
4.10	Delivery Note(s)	
4.11	Discipline Inspection Checklists	

SECTION 5

Section No.	Contents	Applicable Y or X
5.0	**Surface Protection/Coatings**	
5.1	Inspection & Test Plan (ITP)	
5.2	Coating Procedure	
5.3	Coating Inspection Reports	
5.4	Coating Data Sheets	
5.5	Calibration Records – Test Equipment	
5.6	Coating Inspector's Certification	
5.7	Certificate of Conformity	
5.8	Inspection Release Certificate (IRC)	
5.9	Delivery Note(s)	

SECTION 6

Section No.	Contents	Applicable Y or X
6.0	**Proprietary Equipment**	
6.1	Purchase Order (Un-priced)	
6.2	Data Sheet(s)	
6.3	Certificate of Conformity	
	EC Declaration of Conformity (if applicable)	
	EC Declaration of Incorporation (if applicable)	

SECTION 7

Section No.	Contents	Applicable Y or X
7.0	**Works Assembly & Testing**	
7.1	Works Setting to Works Document	
7.2	Quality Plans	
7.3	Cable Test Sheets	
7.4	Factory Acceptance Test Document	
7.5	Calibration Records – Test Equipment	
7.6	Report of Thorough Examination	
7.7	Inspection Release Certificate (IRC)	
7.8	Packing & Delivery Note(s)	

SECTION 8

Section No.	Contents	Applicable Y or X
8.0	**Site Assembly & Testing**	
8.1	Site Setting to Works Document	
8.2	Risk Assessments	
8.3	Method Statements	
8.4	Cable Test Sheets	
8.5	Site Acceptance Test Document	
8.6	Calibration Records – Test Equipment	
8.7	Report of Thorough Examination	

SECTION 9

Section No.	Contents	Applicable Y or X
9.0	**Certification**	
9.1	EC Declaration of Conformity	
9.2	Certificate of Conformity	
9.3	Client Handover Certificate	

TIPS FOR BEST PRACTICES

The LTQR Index's flexibility allows it to be customised to meet the unique needs of each client and project, ensuring both relevance and effectiveness. Consider your project's specific requirements and craft an index that aligns perfectly.

2.3.2 Concurrency in LTQR

Imagine this: You're overseeing a complex, high-stakes project, like building a nuclear reactor or manufacturing critical defence equipment. Every single component needs to be perfect – every weld, every material, every test. Now, picture having access to a system where all the quality data, from the first design sketch to the final inspection, is updated and available in real-time. That's the magic of concurrency in Lifetime Quality Records (LTQR). Concurrency means that quality records aren't just created; they're updated and shared in the moment – simultaneously across all teams involved. Whether you're in the design phase, during assembly, or after maintenance, every change, every inspection, every performance test gets recorded as it happens. No waiting, no outdated information – just an immediate flow of critical data.

For example, imagine a crane being built in your factory. As components are assembled, their performance is logged into the LTQR system on the spot. If any issues arise during testing, they're flagged instantly, and everyone – whether they're on the factory floor or in the engineering office – sees the update right away. This keeps the entire team in sync and able to make quick, informed decisions, minimising delays and mistakes.

In industries like nuclear and defence, where the smallest error can have catastrophic consequences, you can't afford to have isolated teams working with outdated or incomplete information. Concurrency makes sure that every step of a component's lifecycle is connected in real time.

Think of an overhead crane: from the first design concepts, through material sourcing, to assembly and final testing – **every step** is tracked in parallel. If an issue crops up during assembly, it's immediately logged into the LTQR, and that information flows to the testing team, the maintenance team, and anyone else who needs it. This constant, real-time updating ensures that nothing gets missed, and any issues are addressed before they snowball into bigger problems.

When all quality data is flowing in real-time, collaboration becomes seamless. Imagine different teams working across different time zones or even different continents – designers, engineers, quality inspectors, maintenance technicians – all working together, even when they're not physically in the same room.

With concurrent LTQR, all teams access the same up-to-date information, which makes it easier to coordinate and tackle challenges together. Everyone is on the same page, which means fewer miscommunications and fewer errors. Plus, because every action is recorded and timestamped, there's clear accountability – no finger-pointing, just a shared responsibility for ensuring the highest quality at every stage.

The complexity of large-scale projects in industries like nuclear energy or defence can seem overwhelming. You're dealing with long timelines, multiple teams, and complex components that evolve over years. But with concurrency in LTQR, all these moving parts stay aligned and on track.

Let's say your company is working on the construction of a new reactor, and your crane team is responsible for handling critical materials. As the crane is built, the progress is logged in real-time, and as materials arrive for the reactor, their quality checks and certifications are updated in parallel. This means that every part of the project is always in sync, reducing the risk of delays or mismatched components. The concurrent flow of data makes managing complex projects smoother, more efficient, and more predictable.

As technology advances and demands for more complex systems grow, concurrency in LTQR isn't just nice to have – it's essential. With the need to scale up and handle increasingly intricate projects, having a concurrent LTQR system in place gives manufacturers the flexibility to meet future challenges head-on. Whether you're building advanced defence equipment, reactors, or other critical systems, this approach keeps you ahead of the curve.

In a world where safety and reliability are non-negotiable, the concurrency of LTQR is a game-changer. It ensures that quality is never compromised, no matter the complexity of the task or the stage of the project. By making sure that every update, every change, and every inspection is captured in real time, you can prevent costly mistakes, enhance team collaboration, and keep your projects running smoothly. With concurrency, your quality records are always in sync, allowing you to manage the entire lifecycle of your components with confidence and precision.

2.3.3 Completeness: The Full picture

Imagine you're building a puzzle. To get that satisfying "click" when the last piece fits, you need every piece to be present. Completeness in an LTQR is no different. It's about documenting every single detail of a crane's life, from the very first design sketch to the day it's decommissioned. That means:

Accuracy: no room for mistakes

Now, imagine a crane about to lift a critical load – just a slight miscalculation in weight could lead to disaster. The same principle applies to Lifetime Quality Records (LTQR). Every record must be precise because even the tiniest mistake can have major consequences.

Why Accuracy Matters

- **Correct measurements.** Whether it's material specifications, weight tolerance, or part dimensions, every figure must match real-world data. A small deviation could mean the difference between a safe lift and a catastrophic failure.

- **Testing results.** If a crane undergoes a load test, the results should match its actual performance down to the decimal. If it lifted five tons, the LTQR must show five tons – not 4.5 or 5.2. Precision ensures that performance evaluations remain reliable.

- **Material provenance.** Every batch of steel or critical component must have a clear origin and quality certification. LTQR document this, ensuring that inferior materials never make their way into production.

How LTQR Support Data Integrity

- **Reliability and trust:** LTQR serve as the ultimate source of truth in a Quality Management System (QMS). They track everything from the initial design phase to final inspections and maintenance records. Anyone reviewing them should trust that the data reflects reality.

- **No room for guesswork:** Imagine an engineer evaluating whether a crane can handle a heavy-duty lift. If LTQR provide accurate inspection results, they'll know exactly how much stress the crane can endure and where it's most vulnerable. But if the records contain errors? That's a disaster waiting to happen.

2.3.4 LTQR Page Numbering

In the realm of Lifetime Quality Records maintaining the integrity and retrievability of documentation is paramount. A critical aspect of this process is the "drop test" – the ability to seamlessly reassemble an LTQR file or dossier if it becomes disassembled, whether intentionally or accidentally. Central to this capability is the implementation of effective page numbering strategies.

Two widely accepted approaches to LTQR page numbering are:

Method 1: Full Sequential Numbering

This approach involves numbering the entire LTQR file, or each individual section, sequentially from start to finish (e.g. "Page 1 of X"). While this method provides a clear and straightforward numbering system, it is best suited for simple or straightforward LTQR files or sections. It requires the file or section to be fully completed before numbering can be finalised, and any subsequent changes necessitate renumbering the entire document or section.

Method 2: Individual Document Numbering

In this method, each document within the LTQR is numbered independently (e.g. "Page 1 of 4, 2 of 4" etc.). Subsection indices list each document along with the number of pages it contains, and the section index's "number of sheets" column indicates the total number of sheets in each section. This approach allows for modifications to individual documents without the need to renumber the entire section or LTQR, making it particularly suitable for large or complex LTQR.

By adopting a structured page numbering system, organisations can enhance the manageability and reliability of their LTQR, ensuring that documentation remains intact, accessible, and trustworthy throughout the project's lifecycle.

2.3.5 Traceability: finding the origins

Think of traceability as the detective work of the crane world. Every part of the crane has a history, and if something goes wrong, you need to trace it back to its source. Every detail, from materials to labour, should be linked to its original record.

Material traceability. If a weld fails or a part cracks, traceability allows you to find out exactly which batch of material was used, when it was sourced, and who approved it.

Labour traceability. Who assembled the crane? Which team conducted the safety checks? With proper traceability, you can pinpoint who was involved in each stage of the crane's lifecycle.

Component traceability. If a motor breaks down, you'll need to trace its specific part number and batch code to determine its origin. LTQR ensure you can find this information quickly.

Why is this fun? Because it's like following a treasure map! If you uncover a problem, traceability helps you unearth where it came from, how it happened, and most importantly, how to fix it. It's a safety net that protects both crane operators and manufacturers from the unknown.

2.3.6 Accessibility: The Crane's open library

Here's the thing: Having all the right documentation is useless if you can't access it when you need it most. Accessibility in LTQR means that the right people can get to the right information at the right time.

Paperless movement. Today, more and more companies are turning to electronic quality management systems (eQMS) to digitise and store LTQR. This means that the records aren't tucked away in dusty filing cabinets, but rather, stored in a secure online space.

Permissions. While it's important that the records are easy to find, it's equally important to restrict access to unauthorised personnel. Only the right team members – engineers, quality assurance experts, and inspectors – should be able to view and modify certain records.

Real-time updates: Digital LTQR make it easier to update records instantly. If a crane part is replaced or a test result changes, the system can reflect the new information in real-time.

The beauty of accessibility lies in the convenience – whether you're at the crane site or in the office, you can pull up the crane's entire history at a moment's notice. Whether it's a quick fix or a full audit, accessibility is key.

2.3.7 Security and Integrity: Keeping Things Safe

LTQR need security to protect them from unauthorised access, tampering, or loss. This is especially important because LTQR contain sensitive, vital information that directly impacts the safety and operation of cranes.

Digital Security. Passwords, encryption, and secure servers ensure that no one can access or alter the LTQR without proper authorisation.

Physical Security. If some records are kept in physical form, they need to be stored securely in lockable cabinets or rooms with restricted access.

Backup Plans. Regular backups are essential in case records get lost due to technical failures. A good LTQR system will have an automatic backup strategy to prevent data loss.

The security of LTQR ensures that the trust between manufacturers, crane operators, and regulatory bodies remains intact. It keeps the documents untouched and unchanged, preserving the integrity of the crane's history.

2.3.8 Retention: holding onto the past

Finally, we come to retention – and this one is all about time. LTQR don't just serve as temporary records; they're designed to stand the test of time. Different industries and regions have specific regulations on how long these records need to be stored.

Industry regulations. For example, in the nuclear and defence industry, LTQR for overhead cranes might need to be retained for decades, given the potential long-term effects of safety issues.

Legal compliance. Retaining records for the required duration is critical for compliance. If an incident occurs years after a crane was built, having accessible LTQR means that the issue can be traced back and resolved.

Archiving. After a certain period, older LTQR can be archived, but they must remain accessible and retrievable if needed.

Retention isn't just about holding onto paper or data for the sake of it. It's about ensuring that no matter how old the crane, its full history remains available to those who need it. Think of it as a time capsule – preserving the crane's legacy and ensuring safety across generations.

2.4 THE IMPORTANCE OF LTQR

You might wonder, "Why go to all this trouble? Isn't it just another layer of bureaucracy?"

Not at all! LTQR are critical for several reasons:

✓ Ensuring Safety

Cranes often work in hazardous environments, lifting massive loads. A single failure can lead to catastrophic accidents. LTQR ensure that every component has been tested and approved, reducing the risk of failures.

✓ Regulatory Compliance

Industries like nuclear power, aerospace, and heavy manufacturing have strict safety regulations. LTQR provide the documentation needed to prove compliance, showing inspectors that the crane meets all standards.

✓ Operational Efficiency

LTQR help operators understand the crane's history, making it easier to plan maintenance and avoid unexpected breakdowns. For example, if a component was replaced five years ago, the LTQR will show when it might need attention again.

✓ Traceability

If something goes wrong, LTQR allow manufacturers to trace the problem back to its source – whether it's a faulty component or a manufacturing step. This traceability is crucial for solving problems and preventing future issues. Imagine a crane designed to transport and assemble missile components in a secure defence facility.

LTQR would document every aspect of the crane's design, material selection, welding procedures, and load testing. Should any component require replacement or maintenance, the LTQR would provide a complete history, ensuring compatibility and reliability.

✓ **Facilitating Continuous Improvement:** LTQR provide data that can be analysed to identify trends, improve processes, and enhance crane performance over time.

The Role of LTQR across key professions

Lifetime Quality Records are a vital cornerstone of safety, compliance, and efficiency across various professions in highly regulated industries. Their impact extends beyond documentation – LTQR support decision-making, risk management, and operational excellence in roles such as:

Nuclear Safety Case Specialists	Verify compliance with strict safety regulations, ensuring critical systems meet required standards.
Site Inspectors	Assess quality, safety, and adherence to procedures during construction, maintenance, and operational phases.
Project Planning & Control Professionals	Track progress, manage risks, and maintain accountability in complex projects.
Commissioning Engineers	Validate system performance to ensure equipment and processes meet regulatory and operational standards.

Quality Assurance (QA) Teams	Maintain rigorous quality control, ensuring compliance with industry regulations and contractual requirements.

These are just a few examples of how LTQR are integrated into critical professions.

Below is a preview of projected skill areas for the next 10 years, further proving the growing importance of LTQR in ensuring high standards and regulatory excellence in the future workforce. It was predicted in 2017 that these skills would become increasingly important, and now, in 2025, we can clearly see that the demand for these skills is massively growing. This is one of the key reasons I wanted to write this book – to help you stay ahead of the curve and capitalise on the expanding opportunities in these fields.

Fragile skills areas predicted over the next 10 years

- Nuclear Safety Case Specialists
- Control and Instrumentation Engineers
- Site inspection
- Project planning & control
- Commissioning Engineers
- Emergency planning
- Quality Assurance

(Cogent, 2017)

By the time you finish this book, you will be fully equipped to deliver projects with ease and confidently navigate career changes within these industries, opening doors to new opportunities in nuclear, defence, manufacturing, and beyond.

FOOD FOR THOUGHT

The cranes don't just carry the load; they carry the weight of compliance, security, and safety. If something goes wrong, there is no going back. That's why LTQR are vital – they ensure that everything from installation to ongoing maintenance and modifications is meticulously tracked, checked, and certified. Without correct records, you are putting your operations, your team, and the public at risk. So, the next time you see an overhead crane effortlessly moving massive loads, remember: behind that powerful machine is a story meticulously documented in its LTQR.

2.5 DOCUMENTATION STANDARDS

When it comes to Lifetime Quality Records (LTQR), think of documentation standards as the rulebook that everyone follows to ensure everything is neat, clear, and consistent. Picture this: You're assembling a world-class crane, one of the most sophisticated pieces of machinery out there. Would you just toss your instructions, sketches, and test results into a drawer, hoping everything magically falls into place? Absolutely not! You need a structured, organised, and standardised approach to document every step of the crane's journey, from design to operation.

Let's take a deep dive into the best practices for documentation standards, making sure that your LTQR system isn't just a pile of papers but a reliable, consistent, and foolproof record of a crane's life. Ready? Let's get started!

1.   Standardised forms: building consistency

Consistency is key in any complex operation – especially in the world of overhead cranes. Without it, the chance for confusion or error skyrockets. So, how do we ensure that everyone is on the same page? With standardised forms and templates, of course! These forms are like the uniform that all the crane's documentation wears.

Here's how standardised forms work their magic:

> *Uniformity across teams.* Whether it's the design team, the assembly line workers, or the safety inspectors, everyone will use the same form to log information. This ensures that no

matter who's filling out the record, it's in the same format, with the same categories of information.

Clear instructions. Each form has predefined fields that guide the person filling it out. This prevents omissions or confusion. For example, a standardised inspection checklist ensures that the inspector doesn't miss any steps or key components.

Easy comparisons. When you use standardised forms across the board, it's much easier to compare records, spot patterns, and identify areas for improvement. After all, you can't improve what you can't measure – and measuring becomes a lot easier when every record follows the same format.

2. Accuracy and completeness: The foundation of trust

Imagine this scenario: You've just finished a huge crane project. The crane is ready for its first major lift, but there's one small issue – the load charts don't match the final specifications. If only the records had been accurate and complete, you wouldn't be facing this problem.

Here's why accuracy and completeness are absolutely non-negotiable:

Accurate data. Every single number, measurement, and fact has to be spot on. Whether it's the weight capacity of a crane, the specifications of a component, or the results of a safety test – accuracy ensures that the crane performs safely and according to design. A simple mix-up, such as a wrong weight capacity, can have catastrophic consequences.

Complete data. If any part of the process is left undocumented, it could cause major headaches down the road. Imagine there's a part that was replaced during assembly, but the replacement was never recorded. Later, if that part fails, you won't have a record to trace the issue. This is why every activity, whether it's welding a joint or conducting a safety inspection, needs a full description. If something's missing, it's like trying to read a book with pages torn out – it's incomplete, and you'll miss vital information.

So, whether it's a quick test or a major overhaul, complete and accurate records are the difference between smooth sailing and a serious safety problem. These records are trustworthy – they're the backbone of crane safety, reliability, and performance.

3. Organised storage: A library for Cranes

If you've ever been on a scavenger hunt, you know how frustrating it can be to find that one piece of treasure that seems to be hidden in the deepest corners. That's exactly how it feels when your records aren't stored properly. Organised storage isn't just about neatness – it about efficiency, speed, and security.

Here's how to ensure your LTQR are always ready to be found when you need them:

Digital vs. physical. Most modern companies opt for digital storage, which makes organising and retrieving records a breeze. With digital systems, you can easily sort records by date, project, or component type. This makes it so much easier to track a crane's history and even spot recurring issues. Think of it as a giant digital filing cabinet that you can search through instantly.

Folder structure. Whether you're going digital or physical, a logical folder structure is essential. For example, organise records by the crane type, testing phase, or component category. You can then dig down to find a specific record with a few clicks or a simple search.

Version control. If changes or updates are made to a record, it's important to have version control. This keeps track of what changed, when it changed, and who made the change. It ensures that the most up-to-date information is always at hand and prevents confusion.

Let's make an analogy: If a crane's life were a movie, the LTQR would be its script. You wouldn't want a scattered, messy script with missing pages, right? An organised system ensures that every line of the story is in the right place.

TIPS FOR BEST PRACTICES

Use Simple Templates. Simplicity is key! Overcomplicated forms can lead to errors or omissions. Keep templates clean, straightforward, and easy to fill out.

Consistent Terminology. Make sure everyone is using the same terminology across all forms. For example, if you're describing a "safety check", don't switch to calling it a "test" halfway through. Consistent wording helps prevent confusion.

Regular Audits. Just as you'd inspect a crane regularly, the documentation system itself should be audited. Is everyone following the standardised forms? Are records being stored properly? Periodic reviews ensure that the system stays in tip-top shape.

THIS CHAPTER EXPLORES THE importance of LTQR system and highlights how LTQR work within a Quality Management System (QMS) to maintain high standards throughout a crane's lifecycle.

KEY TOPICS COVERED:

- Essential Components of an LTQR System: The role of QMS, document control, audits, training, and technology in maintaining accurate and reliable records.

- LTQR & Quality Management: How LTQR enhance data accuracy, continuous improvement, risk management, and customer confidence.

- ISO 9001 Compliance: The structure, traceability, security, and improvement processes that ISO 9001 brings to LTQR.

- Implementation Strategies: Steps to define objectives, standardise documentation, and integrate LTQR into daily operations effectively.

CHAPTER 3

Effective LTQR System

THE POWER OF LTQR IN QUALITY MANAGEMENT: BUILDING A SYSTEM THAT WORKS

When it comes to cranes – or any complex, high-stakes machinery – quality isn't just a nice-to-have; it's everything. Every nut, bolt, weld, and component must meet the highest standards of safety, reliability, and compliance. That's where Lifetime Quality Records (LTQR) step in. They serve as the ultimate playbook, documenting a machine's journey from design to decommissioning. But LTQR don't work alone. They're part of something much bigger: the Quality Management System (QMS). Together, they form a powerhouse duo, ensuring that every crane (or any product in your industry) meets the gold standard of excellence.

So, let's dive in and explore what makes an LTQR system effective and how it fits into the grand scheme of quality management.

3.1 KEY COMPONENTS OF A SUCCESSFUL LTQR SYSTEM

Think of an LTQR system like a well-oiled machine – it requires key components working together to keep operations running smoothly. Here are the building blocks that make it work:

- **Quality Management System (QMS):** The backbone of any LTQR system. It provides the framework for creating, maintaining, and auditing LTQR, ensuring compliance with industry standards.

- **Standard Operating Procedures (SOPs):** These are the "how-to" guides for everything quality-related. SOPs ensure consistency, accuracy, and proper training of employees handling LTQR.

- **Document Control:** A solid system for version control, approval workflows, and archiving. This keeps records accurate, up-to-date, and tamper-proof.

- **Audit and review mechanisms:** Regular audits (both internal and external) verify that LTQR are complete, compliant, and aligned with best practices.

- **Training and competency:** Employees must be well-versed in quality management, regulatory requirements, and relevant tools to keep LTQR effective and reliable.

- **Technology and tools:** Digital documentation, enterprise resource planning (ERP) systems, and electronic document management systems (EDMS) streamline the LTQR process, improving efficiency and accessibility.

3.2 HOW LTQR SUPERCHARGE YOUR QMS

Imagine a crane as a high-performance sports car. The crane itself is a beast of engineering, built for strength and precision. The QMS? That's the pit crew – working behind the scenes to ensure peak performance. And LTQR? They're the race logs, capturing every test, repair, and performance update to keep that machine running safely and efficiently. By integrating LTQR into a robust QMS, companies can ensure that their cranes (or any other product) are not only built to last but also continuously evolving to meet the highest industry standards. And that's how you build a future of quality, safety, and excellence – one well-documented step at a time.

Let's see how LTQR drive quality within a QMS.

1. Data accuracy & integrity: The Foundation of trust

Quality management is only as good as the data that supports it. If records are inaccurate, outdated, or incomplete, decision-making becomes a guessing game – something you don't want when dealing with cranes lifting massive loads.

✓ **How LTQR help:** They act as a single source of truth, recording everything from design specs to maintenance logs, ensuring that every decision is based on verified data.

2. Continuous improvement: Cranes that keep getting better

In quality management, the goal isn't just to meet standards – it's to raise the bar. LTQR serve as the ultimate feedback loop, allowing teams to analyse past performance and refine processes for even greater efficiency and safety.

✓ **How LTQR help:** They capture trends over time, pinpointing recurring issues, process bottlenecks, and opportunities for innovation.

3. Risk management: Predict & prevent failures before they happen

When heavy machinery is involved, risk management is a must. You don't just react to problems – you anticipate and eliminate them before they escalate.

✓ **How LTQR help:** They document every design flaw, material defect, and operational hazard ever encountered, creating a risk roadmap that helps manufacturers and operators avoid repeating past mistakes.

4. Customer confidence: Trust built on transparency

In competitive industries, customer trust is everything. Clients want assurance that their equipment is safe, reliable, and built to last. LTQR provide tangible proof of quality and compliance, offering peace of mind to customers and stakeholders alike.

✓ **How LTQR help:** They demonstrate a company's commitment to quality control, transparency, and accountability.

3.3 HOW ISO 9001 CONTROLS LIFETIME QUALITY RECORDS (LTQR)

In the crane industry, where safety, precision, and reliability are paramount, maintaining quality standards is not just a goal – it's a necessity. This is where ISO 9001 comes into play. As one of the most widely recognised international standards for quality management systems (QMS), ISO 9001 sets the framework for ensuring products and services meet consistent quality levels. But how does it specifically control and influence Lifetime Quality Records (LTQR)? Let's explore!

What is ISO 9001?

ISO 9001 is an international standard for QMS, developed by the International Organization for Standardization (ISO). It defines the criteria for a quality management system based on several key principles, such as a strong customer focus, process improvement, and evidence-based decision-making.

At its core, ISO 9001 is about ensuring continuous quality improvement through a structured approach. It requires organisations to document their processes, monitor performance, and strive for continuous enhancement – making it a perfect companion to LTQR.

ISO 9001 provides the structure and discipline needed to maintain high-quality LTQR in manufacturing. By ensuring accuracy, traceability, security, and continuous improvement, it turns LTQR from simple records into powerful tools for quality assurance and compliance.

In essence, ISO 9001 doesn't just control LTQR – it elevates them, ensuring every product or service meets the highest standards from design to decommissioning. In the crane industry, where safety and reliability are non-negotiable, this partnership between ISO 9001 and LTQR is what keeps the wheels – and the cranes – moving safely and smoothly.

ISO 9001 and LTQR: The perfect partnership

LTQR are a detailed record of a product's entire lifecycle, from design and manufacturing to operation and maintenance. ISO 9001 provides the structure and controls necessary to ensure these records are accurate, accessible, and reliable. Here's how:

1. Document control and accuracy

ISO 9001 requirement:

ISO 9001 emphasises the importance of maintaining accurate and up-to-date documentation. Section 7.5 of the Standard outlines requirements for creating, updating, and controlling documented information.

Impact on LTQR – This means all records in the LTQR system must be:

Accurate and complete. Every detail – from material certifications to inspection reports – needs to be precise and verified.

Version-controlled. ISO 9001 ensures that only the most recent and approved versions of documents are used, preventing errors from outdated information.

Clearly identified. Each document in an LTQR system must be labelled clearly, with unique identifiers to prevent mix-ups.

2. Traceability and record-keeping

ISO 9001 requirement:

ISO 9001 requires organisations to maintain records that demonstrate compliance with quality processes and standards. It also emphasises traceability, ensuring you can track each record back to its source.

Impact on LTQR – LTQR must link every document – whether it's a material batch record or a maintenance log – to its origin. This traceability means that if a quality issue arises, you can quickly find out where things went wrong and why.

3. Controlled access and security

ISO 9001 requirement:
ISO 9001 mandates that organisations control access to documented information, ensuring only authorised personnel can view or edit critical records.

Impact on LTQR – In the context of LTQR, this means:

Secure storage. Records must be stored in a way that prevents unauthorised access or tampering. Digital LTQR often use encrypted systems with role-based access controls.

Audit trails. Every change made to an LTQR should be logged, creating a clear history of who accessed or modified the records and when.

4. Continuous Improvement and LTQR analysis

ISO 9001 requirement:
A core principle of ISO 9001 is continuous improvement. Organisations must regularly review their QMS and use data to identify areas for enhancement.

Impact on LTQR – LTQR play a crucial role in this process:

Data-driven decisions. By analysing LTQR data, manufacturers can spot patterns or recurring issues and make informed decisions to improve processes.

Feedback Loop. Inspection records and maintenance logs help identify weaknesses in design or production, feeding back into the continuous improvement cycle.

5. Audits and compliance checks

ISO 9001 requirement:

Regular internal and external audits are a key part of ISO 9001 compliance. These audits verify that the organisation is following its QMS procedures and that documented information is accurate and complete.

Impact on LTQR – During an audit, LTQR are often reviewed to ensure they:

> *Meet compliance standards.* Every record must show that the crane was produced and maintained according to industry regulations.

> *Are up-to-date.* Outdated or incomplete LTQR can result in audit findings, so maintaining them consistently is crucial.

3.4 IMPLEMENTATION OF LTQR SYSTEM

Implementing a robust Lifetime Quality Records (LTQR) system is crucial for ensuring that cranes operate safely, efficiently, and in compliance with regulatory standards. A well-structured LTQR system helps capture all relevant data throughout a crane's lifecycle, from design and manufacturing to maintenance and decommissioning. This chapter provides practical guidance on implementing an effective LTQR system in the crane industry, focusing on planning, execution, and continuous improvement.

Understand Your LTQR Objectives

Before implementing an LTQR system, it's essential to clearly define your objectives. Consider what you want to achieve with your LTQR:

- **Compliance:** Ensure that the LTQR system meets all regulatory and industry standards.

- **Safety:** Use LTQR data to minimise risks and enhance safety in crane operations.

- **Quality assurance:** Maintain high-quality standards for crane components and operations.

- **Operational efficiency:** Streamline processes, reduce downtime, and optimise maintenance schedules.

- **Client requirements:** Ensure that LTQR are tailored to meet specific client specifications and contractual obligations.

By understanding your specific objectives, you can tailor your LTQR implementation strategy to meet those needs.

Plan Your LTQR Framework

A successful LTQR system begins with a well-thought-out framework that outlines what records to maintain, how to manage them, and who is responsible for each part of the process.

To ensure records are created, verified, and approved on time, establish clear workflows:

Design workflow:

1. Complete design documentation and internal reviews.
2. Submit for customer and regulatory approvals.
3. Include approved documents in LTQR.

Procurement workflow:

1. Inspect incoming materials and record findings.
2. Match materials to purchase orders and specifications.
3. Archive material certificates in LTQR.

Manufacturing workflow:

1. Track every step of fabrication with detailed records.
2. Conduct inspections at predefined quality checkpoints.
3. Resolve non-conformances and update LTQR with corrective actions.

Testing workflow:

1. Perform functional and load tests according to pre-approved procedures.
2. Document results and cross-reference them with test plans.
3. Have third-party certifications, if required, validated and included.

Installation and handover workflow:

1. Compile all records in an indexed LTQR package.
2. Conduct final inspections and resolve outstanding issues.
3. Present LTQR to the customer for approval.

Establishing LTQR Workflows:

Data collection at each stage

- Design phase → Capture engineering drawings, material specifications, and approval records.

- Manufacturing phase → Document weld inspections, NDT results, and load tests.

- Testing & commissioning → Record performance data, compliance checks, and defect reports.

- Operation & Maintenance → Log preventive maintenance, part replacements, and incident reports.

Assigning responsibilities

- Engineers → Record design specifications and approvals.

- Manufacturing Teams → Maintain production and inspection records.

- Quality Inspectors → Verify testing results and compliance documents.

- Maintenance Personnel → Update service logs and corrective actions.

Ensuring compliance with industry standards

- ISO 9001 → Maintain structured and auditable documentation.

- Regulatory Bodies → Align records with safety and legal requirements.

- Client Contracts → Adapt LTQR to specific customer demands.

Define the scope of LTQR

Determine the scope of LTQR system by identifying which models, components, and processes require documentation. Consider the following:

- **Types of records:** Decide on the types of records to include, such as design documentation, inspection reports, maintenance logs, compliance certificates, and incident reports.

- **Lifecycle phases:** Ensure that records cover all phases of the crane's lifecycle, from design and manufacturing to maintenance, modifications, and decommissioning.

- **Data requirements:** Define the specific data points to capture in each record, such as inspection dates, test results, maintenance actions, and personnel involved.

- **Client-specific requirements:** Incorporate any specific documentation needs or quality criteria required by clients. This may include additional inspections, tests, certifications, or formats as per client contracts or project specifications.

Establish Documentation Standards

Create standardised documentation formats and templates for all LTQR records. Consistency is key to maintaining clear, accurate, and complete records.

- **Use standard templates:** Develop templates for different types of records (e.g. inspection checklists, maintenance logs, incident reports) to ensure uniformity.

- **Define data entry guidelines:** Establish clear guidelines for data entry, including required fields, acceptable formats, and documentation practices.

- **Ensure clarity and detail:** Ensure that all records are detailed, clearly written, and free of ambiguity. Include all relevant information, such as dates, personnel, findings, and actions taken.

- **Adapt to client specifications:** Modify templates and documentation practices to align with client-specific requirements. This may involve creating customised forms, adjusting data fields, or adhering to particular documentation standards requested by the client.

Now that we have laid the groundwork for understanding what LTQR are and why they are critical, we can explore how to design, implement, and maintain LTQR for overhead cranes in the following chapter. In the next chapter, we'll dive into designing the LTQR framework, which involves setting up the organisational structure, deciding what documents and records need to be included, and mapping out workflows for quality control.

I **N THIS CHAPTER, WE** take a deep dive into the process of designing a solid and effective LTQR framework tailored for any product in your business. We break down the essential steps needed to build a comprehensive LTQR system that ensures success throughout the product's entire lifecycle, from design and manufacturing to operation and maintenance.

This chapter guides you through the key elements of creating an LTQR framework that supports consistency, accuracy, and traceability of all records. By focusing on the design, structure, and documentation standards, we ensure that your LTQR system aligns with industry best practices and regulatory requirements. You'll learn how to set up a system that captures critical data, maintains security and integrity, and facilitates easy access for authorised personnel.

By the end of this chapter, you'll be equipped with the tools and knowledge to implement a robust LTQR framework that enhances product quality, ensures compliance, and supports long-term operational success. Let's continue the journey towards mastering LTQR for your business!

CHAPTER 4

Designing YOUR LTQR
(Ideas & Structure)

This chapter takes a personalised approach to LTQR design, helping organisations tailor their Lifetime Quality Records (LTQR) system to fit their specific industry, processes, and compliance needs.

In the nuclear and defence industry, the design of a Lifetime Quality Record (LTQR) framework must address the unique complexities of manufacturing. This chapter provides a step-by-step guide to creating an LTQR framework that ensures compliance, traceability, and efficiency while meeting the stringent demands of the sector.

4.1 DEFINING THE CORE DOCUMENTS IN YOUR LTQR

Every LTQR system must define and standardise the types of documents included.

What Records Should Be Included? *(Tailor based on your needs!)*

- **Design & Engineering Documents** *(CAD drawings, design approvals, load calculations)*

- **Material Traceability & Certifications** *(Mill certificates, supplier records, batch numbers)*

- **Manufacturing & Fabrication Records** *(Welding logs, assembly reports, process documentation)*

- **Testing & Inspection Reports** *(Non-destructive testing (NDT), load testing, functional testing results)*

- **Non-Conformance & Corrective Action Reports** *(Defect reports, root cause analysis, resolutions)*

- **Final Compliance & Certification** *(ISO compliance, regulatory approvals, third-party audits)*

- **Customer-Specific Documents** *(Specialised testing, additional requirements, warranties)*

CUSTOMISATION TIP

If client approval is required at multiple stages, include customer review checkpoints in your LTQR system.

4.2 CHOOSING THE RIGHT LTQR STRUCTURE FOR YOUR BUSINESS

LTQR can be structured in different ways, depending on the company's workflow and priorities.

LTQR Structuring Methods

LTQR approach	Best for	Key focus areas
Project-based LTQR	Custom crane builds, large contracts	Tracks each crane/ project separately
Product Lifecycle LTQR	High-volume manufacturing	Focuses on materials, design, production, and maintenance
Component- based LTQR	Suppliers, spare parts	Tracks individual parts, certifications, and quality

Which Structure Works for You?

Project-Based LTQR – Ideal for businesses that design and build custom cranes, where each project requires unique documentation tailored to specific client needs.

Example: A company constructing bespoke offshore lifting systems would track material traceability, weld records, and load testing for each individual project.

Product Lifecycle LTQR – Essential for companies that track long-term quality performance, providing complete traceability from production to decommissioning.

Example: A nuclear site using overhead cranes for radioactive material handling might maintain detailed records covering initial design, installation, periodic inspections, and end-of-life disposal.

Component-Based LTQR – Best suited for manufacturers producing standardised parts, ensuring consistent quality across multiple production batches.

Example: A manufacturer of crane gearboxes may focus on machining tolerances, batch test results, and supplier certifications to maintain uniform quality.

Digital vs. Manual LTQR systems: Which one is right for you?

Modern LTQR can be stored as paper records, PDFs, spreadsheets, or fully digital databases. Each has advantages and limitations.

Comparison of LTQR storage methods

System type	Pros	Cons
Paper-Based LTQR	Simple, low setup cost	Hard to track, risk of loss/damage, slow audits
Excel or Spreadsheet LTQR	Easy to set up, better than paper	Manual data entry, risk of errors, no version control
Document Management System (DMS)	Secure storage, easy retrieval, audit-friendly	Requires software and employee training
Enterprise Resource Planning (ERP) Systems	Automated, fully integrated with workflows	Expensive, complex setup

Choosing the Right System for Your Business

- Small companies without complex compliance needs → Spreadsheets or simple document control.

- Mid-sized businesses handling client projects and audits → A document management system (DMS).

- Large organisations with high compliance demands → A fully integrated ERP system.

Designing Your LTQR workflow and responsibilities

A successful LTQR system requires clear workflows that define how records are created, reviewed, and stored.

Key Workflow Steps:

1. **Record Creation** → who generates each LTQR document? *(e.g. engineers, QC inspectors, suppliers?)*

2. **Review & Approval** → what approvals are needed before finalising? *(e.g. internal review, third-party validation?)*

3. **Storage & Retrieval** → where and how will documents be stored? *(Paper files, digital cloud, ERP system?)*

4. **Audit & Compliance Checks** → how frequently will records be reviewed for accuracy? *(Annual audits, client inspections?)*

A well-designed LTQR framework is the foundation of quality assurance in the nuclear and defence industries. It ensures compliance, traceability, and transparency throughout the product lifecycle.

Here's a table summarising what should and should not be included in Lifetime Quality Records (LTQR). This table highlights key components to ensure the LTQR remains accurate, relevant, and compliant, while also excluding unnecessary or irrelevant information.

What Should Be Included in LTQR?	What Should NOT Be Included?
Design documentation: Design specifications, reviews, calculations, and approvals.	Unrelated records: E.g. general HR policies or unrelated equipment logs.
Material certifications: Full traceability of raw materials.	Preliminary data: Only include final approved versions of documents
Welding and Inspection records: Weld maps, NDT results, and welder qualifications	Unverified information: Ensure all records are validated and approved by authorised personnel.
Functional testing reports: Results of load and safety system tests.	Personal information: Any personal data unrelated to the crane's operation, like names or employee details.
Customer approvals: Documentation of customer inspections and acceptance.	Personal notes or comments: Informal notes or personal opinions not backed by formal documentation.
Manufacturer details: Information about the crane's manufacturer, including the make, model, and serial number.	Excessive internal communication: Internal emails or memos that do not contribute to the technical or operational aspects of the product.
Calibration records: Evidence of regular calibration of any measuring equipment	Redundant documentation: Duplicated records or information that is already covered in other parts of the LTQR

Modifications and upgrades: Documentation of any modifications, retrofits, or upgrades made to the crane, including technical specifications.	Outdated regulatory requirements: References to obsolete laws or outdated standards that no longer apply to current operational requirements
Surface protection certifications: Certification of the coatings or treatments applied to prevent corrosion or wear, including tests and inspections.	Confidential information: Internal or third-party data that is not relevant to crane operations or safety and could violate confidentiality agreements.

4.3 LTQR MISTAKES AND CHALLENGES

Implementing Lifetime Quality Records (LTQR) is a critical task that ensures compliance, safety, and quality across a product's lifecycle. However, many organisations face challenges during the implementation and management of LTQR, leading to errors that can compromise their effectiveness. This section identifies common mistakes made when managing LTQR and provides strategies for avoiding these pitfalls to achieve a robust and efficient LTQR system.

To gain a deeper understanding of these challenges, I invite you to explore Appendix B, where I've gathered first-hand insights from industry professionals through a series of engaging interviews. These conversations shed light on common pitfalls and offer practical strategies for implementing effective LTQR systems.

Incomplete or inconsistent documentation

One of the most common mistakes in LTQR management is incomplete or inconsistent documentation. Missing data, vague entries, and inconsistent formats can undermine the reliability of LTQR.

How to Avoid:

- Standardise documentation: Use consistent templates and formats for all records. Clearly define required fields, data entry protocols, and acceptable formats to ensure consistency.

- Regular training: Conduct regular training sessions to ensure all personnel involved in LTQR management understand the importance of complete and accurate documentation.

- Checklists and guidelines: Provide checklists and guidelines to help personnel record information consistently. These should be tailored to meet both regulatory standards and client-specific requirements.

Lack of clear roles and responsibilities

When roles and responsibilities for LTQR management are not clearly defined, accountability becomes an issue. This can lead to gaps in documentation, delays, and errors.

How to avoid:

- Assign specific roles: Clearly assign roles and responsibilities for creating, maintaining, reviewing, and auditing LTQR. Each team member should know their duties and the importance of their contributions.

■ Create a responsibility matrix: Develop a responsibility matrix that outlines who is responsible for each part of the LTQR process, from data entry to quality checks and audits.

■ Regular reviews: Conduct regular meetings and reviews to ensure all personnel are fulfilling their roles and to address any issues of non-compliance or confusion.

Ignoring client-specific requirements

Failing to customise LTQR according to client specifications is another common mistake. This oversight can lead to dissatisfaction, contract disputes, or penalties.

How to avoid:

■ Engage with clients early: Involve clients from the beginning to understand their specific requirements for LTQR documentation, formats, and content.

■ Customise templates: Develop customisable templates that can be adapted to meet client-specific needs while maintaining compliance with internal and regulatory standards.

■ Review and update regularly: Regularly review and update LTQR practices to ensure they continue to meet evolving client requirements.

Inefficient use of technology

Many organisations fail to leverage technology effectively, resulting in manual errors, inefficiencies, and poor data management.

How to avoid:

- Invest in digital tools: Use digital tools and software to automate data entry, storage, and retrieval processes. Digital systems can reduce manual errors and improve efficiency.

- Regularly update technology: Keep software and tools up-to-date to leverage new features and maintain compatibility with other systems.

Poor record maintenance and storage

Improper storage of LTQR, such as using outdated filing methods or neglecting digital backups, can result in lost or inaccessible records, especially during audits or inspections.

How to avoid:

- Use centralised digital repositories: Store LTQR in a centralised digital repository that is accessible to authorised personnel, with adequate backup solutions.

- Implement regular backups: Conduct regular backups of all LTQR data to prevent data loss due to system failures or cyber-attacks.

- Archive old records securely: Develop a policy for securely archiving old records, ensuring that they are stored in compliance with regulations and are easily retrievable.

Inadequate training and awareness

Lack of proper training and awareness among staff is a frequent mistake that leads to poor LTQR management practices, incomplete records, and non-compliance.

How to avoid:

■ Develop a comprehensive training programme: Ensure all personnel involved in LTQR management receive thorough training on the procedures, standards, and technology used.

■ Conduct regular refresher courses: Offer periodic training sessions to reinforce best practices, update staff on regulatory changes, and introduce new tools or processes.

■ Promote continuous learning: Encourage a culture of continuous learning by providing access to resources, workshops, and industry seminars.

Failing to regularly audit LTQR

Without regular audits, it is challenging to identify errors, discrepancies, or non-compliance issues in LTQR management. This can lead to major problems during regulatory inspections or client audits.

How to avoid:

■ Schedule regular internal audits: Conduct regular internal audits to check the accuracy, completeness, and compliance of LTQR. Use these audits to identify gaps or weaknesses.

■ Use audit findings for improvement: Use audit findings to implement corrective actions, refine processes, and enhance training programmes.

Underestimating the importance of LTQR

Some organisations underestimate the importance of LTQR, treating them as mere paperwork rather than critical tools for ensuring quality, safety, and compliance.

How to avoid:

■ Foster a culture of Quality: Promote a culture that values quality and compliance across all levels of the organisation.

■ Communicate the benefits: Regularly communicate the benefits of effective LTQR management to all staff, emphasising its role in reducing risks, improving safety, and enhancing client satisfaction.

■ Involve leadership: Ensure that senior management is actively involved in LTQR oversight and demonstrates a commitment to maintaining high standards.

Challenges in implementing LTQR for the nuclear and defence sectors

Implementing and maintaining Lifetime Quality Records (LTQR) in the nuclear and defence sectors is no small feat. These industries demand the highest levels of precision, compliance, and documentation due to the critical safety and operational requirements. Below, we dive into the key challenges and provide actionable solutions to address them effectively.

Challenge 1: Managing high volume and complexity of records

Why It's a Challenge.
Overlap or duplication of documents

Difficulty in retrieving specific records during audits or inspections.

Risk of losing critical information in poorly managed systems.

Solutions:

1. Implement digital tools: Use Quality Management Systems (QMS) or document control software to centralise and organise records.

2. Standardise formats: Develop consistent templates for all records to ensure uniformity and ease of use.

3. Categorise records: Divide records into logical sections (e.g. design, manufacturing, testing) for better organisation and retrieval.

4. Audit trails: Maintain version control and metadata to track changes, approvals, and access history.

Challenge 2: Complying with stringent standards

Why It's a Challenge.
Standards often require highly specific and detailed documentation.

Multiple standards may apply simultaneously, leading to over-lapping or conflicting requirements.

Staying updated with changes in regulations can be challenging.

Solutions:

1. Develop a compliance matrix: Map out all applicable standards and cross-reference them with your LTQR framework to ensure all requirements are covered.

2. Assign a compliance officer: Designate a dedicated professional to monitor regulations, ensure updates are implemented, and communicate requirements to the team.

3. Regular training: Train staff on regulatory requirements and changes to ensure all processes align with the latest standards.

4. Third-party audits: Use external experts to review your LTQR system and identify compliance gaps.

Challenge 3: Ensuring traceability and accountability

Traceability is critical in the nuclear and defence industries, where every component, weld, and test must be linked to specific materials, processes, and personnel. Any lapse in traceability could result in regulatory non-compliance or operational failure.

Why It's a Challenge.

Complex supply chains make it difficult to track material origins.

Poor documentation practices can break traceability links.

Human errors, such as missing signatures or mislabelled documents, compromise accountability.

Solutions:

1. Unique identifiers: Assign unique IDs to materials, components, and processes for easy tracking. Use QR codes or RFID for automation.

2. Integrated systems: Use an Enterprise Resource Planning (ERP) system to link procurement, production, and quality records.

3. Process checkpoints: Implement mandatory review and sign-off points to ensure records are complete and traceable.

4. Continuous monitoring: Use real-time dashboards to track the flow of materials and documentation through the LTQR system.

Challenge 4: Resistance to Change

Shifting to an LTQR-based approach often requires significant changes in workflows, technology, and employee behaviour. Resistance from employees or stakeholders can hinder successful implementation.

Why It's a Challenge.

Employees may feel overwhelmed by new documentation requirements or digital tools.

Existing workflows may not align with LTQR processes, causing confusion.

Management may hesitate to invest in new systems or training programmes.

Solutions:

1. Engage early: Involve employees and stakeholders in the planning process to get their buy-in and feedback.

2. Communicate benefits: Highlight the advantages of LTQR, such as improved efficiency, compliance readiness, and reduced risk.

3. Provide training: Offer hands-on training sessions for staff to familiarise them with LTQR tools and processes.

Challenge 5: Balancing thoroughness and efficiency

While LTQR must be comprehensive, overloading it with unnecessary data can reduce its usability. Striking the right balance between thoroughness and efficiency is a significant challenge.

Why It's a Challenge.

Overly detailed records slow down audits and reviews.

Irrelevant information can obscure critical data.

Pressure to include "everything" leads to inefficiencies.

Solutions:

1. Define relevance criteria: Develop guidelines for what should and should not be included in LTQR. Focus on records that are critical for compliance, traceability, and decision-making.

2. Lean documentation: Streamline records to include only essential information, supported by cross-references to avoid duplication.

3. Periodic reviews: Regularly audit the LTQR system to remove outdated or redundant records.

Challenge 6: Long-Term Retention and Security

Overhead cranes in the nuclear industry have long lifecycles, often exceeding 20–30 years. LTQR must be retained for the entire lifecycle and beyond, requiring robust systems for secure and reliable storage.

Why It's a Challenge.
Physical records are prone to damage, loss, or misfiling.

Digital records are vulnerable to cybersecurity threats or obsolescence of file formats.

Long-term retention increases storage costs and complexity.

Solutions:

1. Digital archiving: Transition to digital records with secure, cloud-based storage solutions.

2. Redundancy: Maintain backups in multiple locations, including off-site facilities.

3. Access control: Limit access to sensitive records and use encryption to enhance security.

4.4 LTQR COMPLETION AND SUBMISSION –
THE GRAND FINALE!

Imagine you're assembling the most important jigsaw puzzle of your career. Every piece is crucial, every detail matters, and at the end of the process, you step back to admire the big picture: your product's complete story. Welcome to LTQR Completion and Submission, where every tiny bit of data you've collected comes together to form the final masterpiece – the "Life Time Quality Record" (LTQR). It's like finishing an intricate project, and now it's time to present your work to the world.

In this section, I'll walk you through how to wrap up your LTQR in style, make sure it's flawless, and submit it like a pro.

1. The LTQR checklist: Double-checking the essentials

Before you even think about hitting "submit", you need to ensure that everything is in place – this is your final opportunity to double-check every detail. Think of it as making sure all your ducks are in a row...and trust me, there are a lot of ducks!

- **Documents & records.** Check if all documents are accounted for – dimensional records, calibration reports, surface protection certifications, inspection results, and maintenance logs. Think of these as the pages of a photo album, each capturing a key moment in your product lifecycle.

- **Dates & time stamps.** Make sure all records have the proper dates. Nothing screams "sloppy work" like a mismatch between when something happened and when it was recorded.

- **Signatures & approvals.** Does everything have the right sign-offs? From engineers to quality inspectors, make sure the necessary personnel have reviewed and approved each record.

- **Compliance confirmation.** Are all your records meeting the necessary regulatory standards and industry guidelines? This is your golden ticket to proving your product meets or exceeds safety and operational standards.

2. Quality Check: The Eagle Eye

You've built this masterpiece over months or even years, but before you submit, it's time for the final inspection. You need to be the eagle-eyed perfectionist who catches every detail. Don't just glance through it – go through the records with the intensity of an artist putting the final brushstroke on their canvas.

- **Consistency is key.** Ensure that all data entries match up across records. If one document mentions a part being replaced, verify that it's documented elsewhere in the LTQR, and the part's specifications are correct.

- **Look for gaps.** Any missing information? Are there blank spaces where critical information should be? Those gaps are like missing pieces in your puzzle – no one likes a puzzle with pieces left out.

- **Formatting and organisation:** This is where you can let your inner librarian shine. Is everything easy to navigate? Are the files logically organised and clearly labelled? Clear section breaks, well-structured indexes, and easily identifiable headers will make the review process a breeze.

3. Electronic submissions: The Future is Now

Let's face it – paper records are so last century. While some may still use physical submissions, digital submissions are where the future is headed. Not only are they easier to store, manage, and update, but they're also quicker to submit and can be reviewed with a click of a button.

- **Organise files efficiently.** Make sure your LTQR is in a format that's easy for anyone to open and read. PDF, Excel, or specialised LTQR software – whatever you choose, ensure that it's well-organised, searchable, and not a tangled mess of documents.

- **Use digital tools to check for completeness.** Many modern LTQR software systems have built-in checks to verify if all necessary documents are present and up to date. These tools are like your personal assistant, making sure nothing gets missed.

- **Backing up your submission.** Always keep a backup, because, you know, life happens. Whether it's digital or physical, ensure there's an easily accessible backup, just in case a server crashes or you spill coffee on your laptop (we've all been there).

4. Final review: Time to celebrate

Once you've double-checked everything, gotten your team's approval, and polished the details, it's time to submit your LTQR. It's like finishing a long journey – you've gathered every piece of data, solved every puzzle, and now it's time to present it to the world. This is the grand finale, and you should be proud.

- **Submission Time.** Whether you're submitting to a regulatory body, an internal quality control team, or a client, ensure you submit everything on time. Late submissions can cause delays in approvals or worse, penalties. Be punctual, be professional.

- **Confirmation and acknowledgment:** After submission, always seek confirmation. Did they receive the files? Are they complete? Are there any follow-up actions? This final step ensures that everything is in the right hands and you can kick back, knowing you've done your job well.

5. What Happens After Submission?

Once the LTQR is submitted, don't just sit back and relax (yet). It's time for the official review process, which may involve several rounds of feedback, questions, or additional documentation requests. Don't worry though, this is just part of the process. Here's what to expect:

- **Review.** Regulators, clients, or internal quality control teams will review your LTQR. They may ask questions, request clarification, or even suggest improvements. Be ready to respond with confidence and clarity.

- **Continuous updates.** An LTQR isn't a one-time thing. It's a living document that evolves throughout the lifecycle of the product. As maintenance occurs, upgrades are made, and inspections are carried out, new records will need to be added. Consider your LTQR as an ongoing project that grows over time.

- **Feedback Loop.** If there are issues with the LTQR (perhaps a missing signature or unclear record), be proactive in addressing them. The more responsive and detailed you are, the smoother the process will go.

6. The joy of completion

Congratulations! You've navigated through the labyrinth of LTQR completion and submission. The product is now officially documented, certified, and ready for whatever comes its way – whether it's regulatory inspections, audits, or simply to keep track of its quality journey.

The best part? With a well-organised LTQR in hand, you've not only ensured the safety and reliability of the product, but you've also contributed to a culture of quality and accountability. You've proven that your product's life story is as solid as the steel it's made of.

Now, go ahead – take a well-deserved break. You've earned it! Just make sure to stay on top of updates and future submissions to keep that LTQR in tip-top shape. After all, the story of a product's life is never really finished – it just keeps on going, one quality record at a time.

A TOUCH OF EXCITEMENT

To gain a deeper understanding of LTQR's importance, check out my interview with industry experts in Appendix B. Their insights will help you see what's happening in the industry and why a strong LTQR framework is crucial for success.

MPLEMENTING LTQR IS ESSENTIAL not only for manufacturing processes but also for maintaining comprehensive maintenance records. In this chapter, we explore how LTQR enable swift and effective responses during emergencies. By maintaining well-organised, accessible, and up-to-date records, you can minimise downtime, mitigate risks, and ensure the safety of all involved.

Manufacturing and maintenance records are fundamentally interconnected, working together to uphold product integrity and operational excellence.

CHAPTER 5

The Power of LTQR in crisis management

When the unthinkable happens – whether it's a mechanical failure, an operational mistake, or an emergency situation where every second counts – having a solid LTQR in place can be the difference between a quick resolution and a catastrophic event. Cranes used in the nuclear industry are critical assets, and their failures can pose severe risks, from safety hazards to regulatory violations. That's why they aren't just vital for daily operations; they're crucial during a crisis.

5.1 QUICK ACCESS TO RECORDS IN EMERGENCIES: A LIFESAVER

In the event of a crane breakdown or malfunction, time is of the essence. The longer it takes to identify the problem, the greater the potential for escalation – especially in a high-stakes environment like a nuclear facility. This is where LTQR come into play. They provide an instant, easily accessible record of every detail related to the crane's history, design, maintenance, and repairs. This access can be a lifesaver when you're racing against the clock to restore operations safely.

Imagine you're in a situation where an overhead crane unexpectedly fails, and you need to determine if the cause is related to a recent modification, a known issue with certain components, or perhaps a failure during testing that was overlooked. With LTQR at your disposal, you can:

- Quickly retrieve the relevant inspection reports from the last maintenance cycle.

- Identify recent modifications to the crane that could be contributing to the problem.

- Cross-reference parts and materials used in the last repair to check for any potential faults.

Without LTQR, you'd be left scrambling to find documents, asking around for maintenance logs, or worse – guessing at the cause of the issue. That confusion costs time, money, and, potentially, lives. With LTQR, you can immediately consult the records, pinpoint the problem, and begin corrective action without delay.

How LTQR are your best defence during a breakdown.
During a breakdown, every minute counts. The faster you can identify the root cause and take corrective action, the faster the crane will be back in service. But it's not just about fixing the crane; it's about ensuring that it's fixed correctly and safely. LTQR are your best defence in ensuring this happens effectively.

Scenario: Breakdown during Nuclear facility operations
Imagine that an overhead crane is in use to transport nuclear waste for disposal, and suddenly it stops functioning in the middle of the process. This crane failure can halt the operation, compromise safety, and lead to costly delays. Here's how LTQR come to the rescue:

1. Identify Previous Issues: The LTQR should include all past maintenance records, repairs, and inspections. By reviewing these records, you can quickly identify if the failure is related to a recurring issue or a part that was recently replaced.

2. Detailed Repair History: Each repair and replacement part will be logged in the LTQR. This gives you a detailed history of every component of the crane. If a component is faulty, you'll know when it was last replaced, which vendor provided it, and whether it's under warranty.

3. Manufacturer and Regulatory Documentation: In the nuclear industry, regulatory compliance is non-negotiable. LTQR store all vendor certifications, component test results, and compliance documentation in one place, allowing you to prove that your crane meets the required safety and quality standards during an emergency.

In essence, LTQR allow maintenance teams to make informed, quick decisions, providing a blueprint for how to fix the problem. By ensuring that all repairs are done based on accurate, comprehensive data, you minimise the risk of further failures, compliance issues, or safety breaches.

5.2 LESSONS LEARNED FROM NUCLEAR CRANES IN CRISIS

There's no substitute for real-world lessons, and the nuclear industry has experienced its share of challenges when it comes to crane failures. Some lessons learned underscore the critical importance of LTQR in crisis management:

Case Study 1: Crane malfunction during a reactor shutdown

In one instance, a crane used to move reactor components experienced a failure during a routine shutdown operation at a nuclear facility. Initial attempts to troubleshoot the issue were slow because the operators didn't have ready access to the crane's maintenance and modification history.

Lesson learned:

When LTQR were finally pulled, the history revealed that a critical part of the crane had been replaced only a few months prior – something that hadn't been flagged during regular inspections. The delay in diagnosing the issue added unnecessary time to the reactor shutdown process, and could have caused even more extensive damage.

Solution:

From that point forward, the facility ensured that all LTQR were digitised and accessible via a central database, so operators and engineers could access them immediately in case of an emergency.

Case Study 2: Emergency repair during a critical lift

In another case, an overhead crane experienced a failure during a critical lift of radioactive materials. The team struggled to locate key records to understand the crane's operational status, recent repairs, and modifications, which delayed repairs. By the time the issue was resolved, there was a significant disruption to operations, costing the plant both time and resources.

Lesson learned:

In high-risk environments like this, time lost equals more than just money – it can mean compromised safety. Had the team had immediate access to the crane's complete LTQR, they would have been able to identify the issue much faster, minimising downtime and potential safety hazards.

Solution:

After this incident, the plant implemented real-time tracking of LTQR, ensuring that all historical data (including modification and maintenance records) were not only readily available but also reviewed during regular operational meetings.

The crucial role of LTQR in avoiding future crises
Crisis situations are always unpredictable, but what isn't unpredictable is the response you can have with the right information. By maintaining up-to-date, accurate, and organised LTQR, you're effectively building a data-driven defence against future breakdowns, non-compliance, and safety issues.

The key takeaways for crisis management are clear:

- **Be Proactive, Not Reactive:** Use LTQR to track potential issues before they become crises. Properly maintained records allow you to identify warning signs early and prevent serious breakdowns.

- **Ensure quick, informed decisions:** In an emergency, you can't afford to make decisions without accurate, accessible records. LTQR give you the data to act fast and minimise risks.

- **Don't wait for a crisis:** Build your LTQR system with crisis situations in mind. The more thorough, organised, and accessible your LTQR are, the better prepared you'll be to handle anything that comes your way.

In conclusion, LTQR are not just documentation; they're lifelines. In crisis situations, they give operators, engineers, and safety personnel the information they need to take immediate action, fix problems quickly, and ensure that operations continue safely and efficiently. Whether you're dealing with a minor breakdown or a major operational crisis, LTQR provide the foundation for effective, informed decision-making.

CHAPTER 6

CHAPTER 6

Next Steps

NEXT STEPS: EMBRACING EXCELLENCE THROUGH LTQR

Congratulations! You've just navigated through the exciting world of Lifetime Quality Records (LTQR), and now it's time to roll up your sleeves and get into action. But where do you start? Don't worry, I've got you covered. Think of this as your treasure map to LTQR success, with each step leading you closer to quality and excellence. Ready? Let's dive in!

1. Assess your current processes: The starting line

Before you begin your LTQR journey, it's like checking the map to see where you are. Evaluate your current processes to uncover hidden treasures – or gaps, inefficiencies, and areas for improvement. Knowing your starting point will help you chart the best path forward and figure out what needs tweaking.

2. Set clear objectives: your guiding star

Every adventure needs a goal! Are you aiming to enhance traceability, ensure compliance, reduce risks, or spark continuous improvement? Defining clear objectives will act as your guiding star, helping you navigate through your LTQR journey. These goals will also help you measure your progress and celebrate those victories along the way!

3. Engage your team: Teamwork makes the dream work!

LTQR is a team sport. Whether it's engineers, managers, or the person who keeps the coffee flowing, every player has a role in maintaining quality records. So, gather your crew! Communicate the importance of LTQR, share the vision, and make sure everyone is on board. A little training can go a long way in making sure everyone knows their part in this grand adventure!

4. Develop or refine your LTQR framework: The Blueprint of Success

Now, it's time to build your LTQR framework – think of this as your project's blueprint for success. Use templates, best practices, and the knowledge you've gained here to create or fine-tune your LTQR process. Make sure it's tailored to your specific needs and aligned with industry standards. It's like crafting the perfect set of tools for your LTQR toolbox.

5. Start small and scale up: Rome wasn't built in a day

Every hero's journey begins with one small step – so start small! Apply LTQR practices to a single project or product line. Think of it like trying out a new recipe before you host a grand dinner party. Test, learn, gather feedback, and adjust as needed. Once you've got the hang of it, expand to more areas of your organisation. You'll soon find that scaling up is as easy as pie.

6. Leverage digital tools: The magic wand

Here's where the magic happens! Explore digital tools and software solutions that can automate and streamline your LTQR processes. Imagine having a digital assistant that organises your records, improves accuracy, and keeps everything secure. These tools make it easier to manage and retrieve quality data, and they're your secret weapon in the LTQR world.

7. Continuously review and improve: The Quest for Excellence

Remember, LTQR isn't a one-and-done task – it's an ongoing adventure! Like any great explorer, you'll need to continuously review your quality records, assess their effectiveness, and look for opportunities to improve. Regular check-ins will help you stay on course, ensuring your LTQR practices are always evolving and improving. Continuous improvement is the secret ingredient in maintaining top-notch quality!

8. Stay informed: Stay ahead of the curve

The world is constantly changing, and so is LTQR! Stay updated on the latest industry trends, regulatory changes, and new technologies that might impact your LTQR practices. Attend events, join forums, or even get involved in discussions to keep your LTQR skills sharp and stay ahead of the competition.

Final Thoughts

As we wrap up this exploration into Lifetime Quality Records (LTQR), it's clear that these records aren't just a set of documents – they're the foundation of quality, safety, and efficiency in overhead crane manufacturing. Implementing LTQR is like adding a safety net, ensuring that every aspect of your crane's lifecycle is meticulously documented, from design and engineering to manufacturing, testing, and maintenance.

Think of LTQR as a roadmap that keeps you on track. It ensures that each phase of the crane's journey is thoroughly checked, verified, and perfected. Whether it's ensuring compliance with industry standards or tracking the smallest design detail, LTQR is your most reliable tool for building cranes that are not only top-performing but also safe and compliant. It's about creating something that lasts, something you can trust to perform under pressure.

Remember, quality isn't just about checking boxes – it's about continuously striving for the best. And with LTQR, you're not just documenting a process; you're building a legacy of excellence.

Good luck on your LTQR journey!

My Passion for LTQR

Life Time Quality Records (LTQR) are not just a collection of documents to me – they are the heartbeat of a commitment to safety, quality, and continuous improvement in the crane industry. My passion for LTQR is deeply rooted in the belief that well-documented, thorough processes lead to higher standards of machinery reliability, safety, and operational efficiency. These records are the unsung heroes that keep the wheels of the crane industry turning, ensuring we meet and exceed expectations every step of the way.

THE SPARK OF MY PASSION

My fascination with LTQR started early in my career when I realised how the intricate system of documentation holds everything together. It wasn't just about keeping files in order; it was about creating a culture of accountability, quality, and safety that stretches across a crane's entire lifecycle. Seeing how each step – whether it's design, assembly, or maintenance – plays a part in ensuring the safety of operators and the environment, I realised that LTQR is much more than a requirement. It's the foundation for creating a reliable, trustworthy, and innovative industry.

WHY LTQR MATTERS TO ME

1. A commitment to Quality

LTQR is the embodiment of excellence. Every document, every detail, every step of the way contributes to the bigger picture of quality control. My passion lies in maintaining these high standards from the drawing board to the crane's final operational days.

2. Championing safety

The safety of crane operators and everyone in their environment is paramount. Knowing that LTQR is integral to this mission fuels my determination to advocate for its rigorous implementation.

3. Driving innovation

LTQR isn't just about holding onto tradition; it's about pushing the boundaries of what's possible. By documenting design changes, performance evaluations, and testing, LTQR helps us learn from the past and innovate for the future. I am driven by the potential of LTQR to help shape the next generation in this field.

4. Ensuring accountability

A well-maintained LTQR makes every aspect of a crane's journey traceable. I'm passionate about this transparency and the clarity it brings to the entire industry.

THE CHALLENGES AND REWARDS

Of course, working with LTQR is not without its challenges. The attention to detail required, the constant changes in industry regulations, and the sheer volume of documentation can feel overwhelming at times. But these challenges are what make the work so rewarding. Every challenge is an opportunity to improve the system, to evolve the practices, and to contribute to the growth of the industry.

The sense of pride when I see a meticulously organised LTQR package that not only meets all regulatory requirements but also serves its intended purpose is unmatched. It's not just a job well done; it's a contribution to a culture of excellence, safety, and continuous improvement.

SHARING THE PASSION

One of the most gratifying aspects of my passion for LTQR is sharing it with others. Whether through training sessions, workshops, or collaborative efforts, helping others understand the importance of LTQR is incredibly fulfilling.

LOOKING TO THE FUTURE

As technology advances and industry standards continue to evolve, the importance of LTQR will only grow. My passion motivates me to stay ahead of these developments, to contribute to shaping the future of documentation in the industry, and to ensure that safety, quality, and efficiency remain at the heart of our work. I also aim to inspire future generations to pursue careers in the nuclear and defence sectors, where they can be part of ground-breaking projects.

APPENDIX

APPENDIX A

Journey through the history of overhead cranes

**ANCIENT INGENUITY: THE DAWN OF CRANES
(500 BC–100 AD)**

Imagine Ancient Greece, a land of philosophers, warriors, and architectural wonders. As the grand temples of the Acropolis began to take shape, a curious question emerged – how were these colossal stones, weighing several tons, lifted and placed with such precision? The answer lies in one of humanity's earliest engineering marvels: the first cranes.

The ancient Greeks, known for their ingenuity, devised rudimentary cranes using wooden beams, pulleys, and ropes. These early cranes were powered by human effort or by donkeys walking in

large treadmills, a clever mechanism that amplified their lifting capacity. With this innovation, tasks that once required the labour of hundreds became achievable with far fewer hands.

MEDIEVAL MARVELS (5TH–15TH CENTURY)

In medieval Europe, castles and cathedrals touched the skies, thanks to the next generation of cranes. The "treadwheel crane" became a superstar – picture a giant hamster wheel for humans! Workers inside these wheels could lift heavy stones with ease, making lofty Gothic cathedrals possible. These cranes were often built right into structures, and some medieval treadwheels still survive today!

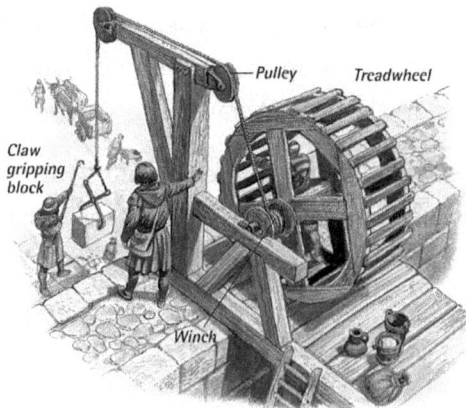

INDUSTRIAL REVOLUTION SPARKS CHANGE (1760–1840)

Fast forward to the 19th century – the Industrial Revolution. Factories hummed, steam engines roared, and cranes got a power boost! No longer reliant on human strength alone, steam-powered cranes could lift heavier loads faster. One of them was constructed by Mr Thomas Smith, of Rodley, near Leeds, for use in a steel works, to lift, lower, and travel with loads up to 15 tons. The crane was designed for hoisting and lowering while travelling transversely or longitudinally, and all the movements were readily controlled from the cage, which was placed at one end of and underneath the transverse beams, and from which the load could readily be seen. All the gear wheels were of steel and had double helical teeth; the shafts were also of steel, and the principal bearings were adjustable and bushed with hard gun metal. This crane had a separate pair of engines for each motion, which were supplied with steam by the multitubular boiler placed in the cage as shown.

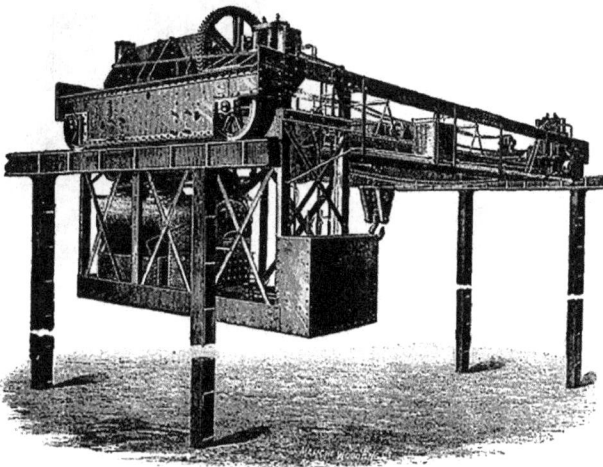

ELECTRIFYING PROGRESS (LATE 19TH–EARLY 20TH CENTURY)

The next game-changer? Electricity! By the late 1800s, electric motors transformed cranes. Overhead cranes became essential in steel mills, power plants, and railroads. Companies like Demag in Germany and Konecranes in Finland started refining designs and setting global standards. These electric cranes weren't just strong – they were smarter, with controls that allowed precise movements.

1914: Electrically operated crane by a control pendant and from an operator cabin attached to the crane.

POST-WAR INNOVATIONS (1945–1980S)

After WWII, the world rebuilt, and overhead cranes evolved with it. New materials like high-strength steel made cranes sturdier and lighter. Hydraulics added power, while gears improved control. Cranes became more specialised, entering shipbuilding, automotive, and even aerospace industries. Ever wonder how those giant aircraft parts are moved? Yes, overhead cranes!

The post-war boom led to widespread adoption of overhead cranes across various industries. Innovations included remote controls, which enhanced precision and efficiency, and improved safety features such as emergency stop systems and load simple sensors. These advancements facilitated mass production, streamlined material handling, and boosted overall industrial productivity.

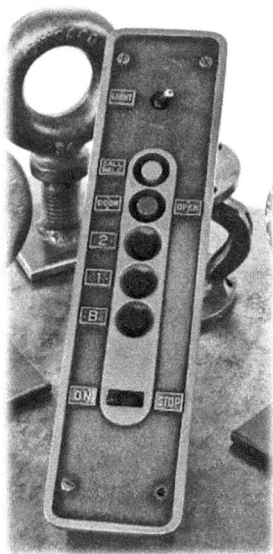

The remote control in early days

DIGITAL DOMINATION (1990S–TODAY)

Step into the world of tomorrow, where overhead cranes aren't just machines – they're technological marvels. Gone are the days of purely manual operation. Today's cranes are powered by cutting-edge advancements that make them safer, smarter, and more efficient than ever before.

Imagine a crane that can think ahead, predict potential maintenance needs, and operate with pinpoint precision – all while being controlled remotely or functioning autonomously. That's not science fiction; it's the reality in high-stakes environments like nuclear power plants and defence facilities, where every movement matters.

Modern cranes are equipped with laser-accurate positioning systems, ensuring that even the most delicate operations – like handling nuclear reactor components or assembling critical defence equipment – are executed flawlessly. Real-time monitoring systems continuously check for anomalies, keeping safety at the forefront.

Leaders in Innovation: Kinetic Solutions Group (KSG/SCX)

Companies like KSG/SCX are at the forefront of this revolution. They design cranes with smart features that redefine what's possible in critical industries, including nuclear power and defence. Their innovations ensure that heavy lifting meets heavy thinking, making operations safer, more efficient, and future-ready.

SCX Overhead Crane

Overhead cranes have come a long way – from human-powered wheels to digital giants. They embody our endless drive to build bigger, smarter, and safer. So, next time you see an overhead crane, remember: it's not just lifting metal; it's lifting centuries of innovation and human ambition!

APPENDIX B

Interviews

INSIGHTS FROM INDUSTRY EXPERTS IN THE NUCLEAR AND DEFENCE SECTOR

1. Interview with Prof Amanda McKay.

Could you introduce yourself and describe your role in the nuclear or defence industry?

I am Prof Amanda McKay, currently Quality lead for the UK Geological Disposal Facility project. Previous roles have been as Director Quality for AWE, Quality and Assurance Director for BBV on HS2. I have over 35 years' experience in working in Quality Assurance, Health and Safety within the Nuclear, Defence, Renewables and oil and gas major projects.

What initially attracted you to the field of quality?

Initially I did not want to transfer to a Quality role from operations but soon discovered that Quality gave me a great overview of how businesses operate and how to improve performance; I've been hooked since.

How long have you been working with quality management systems (QMS) and Lifetime Quality Records (LTQR)?

Five years with Quality systems and just over 20 with LTQR.

What are the biggest changes you've witnessed in quality management practices over the years?

The key things for me are the move away from inspection and quality control to a quality assurance approach, prevention and improvement over detection and correction. In my early days in oil and gas it was always about post work inspection, now we are risk driven and seek to identify issues in planning and design to prevent defects and deliver right first time. Other key changes are the professionalisation of the Quality profession and the move to digital tools and methods.

In your experience, what is the most challenging part of maintaining LTQR?

The key issue with maintaining LTQR is ensuring they are kept current and accessible – when changes occur to facilities and equipment, updating the LTQR is often forgotten about and sometimes the LTQR is stored so it's not readily accessible.

How do you address these challenges on a day-to-day basis?

Ensuring that regular reviews of the LTQR are conducted as part of regular inspections and audits to ensure they are stored correctly, maintained in a current state and reflect the current configuration of the facility. They also need to be included in quality plans for all upgrades and changes. Compiling and storing them electronically makes it easier to comply.

How have LTQR evolved during your career, and what changes have you noticed in their importance or complexity?

Early in my career (1990s) LTQR were very simple records of as built construction/fabrication, very little importance was placed upon them and some were delivered years after commissioning. Today we see them compiled electronically or built into the digital twin of the facility, they are seen as key records and need to be compiled concurrently with the works. The level of detail and types of information are now much greater with records of qualifications of those involved in the works, details of the supply chain and origin of materials etc.

Do you think companies are placing more emphasis on LTQR now than in the past? Why or why not?

Yes, there is greater emphasis on LTQR now, we often say that we deliver the project twice now, once in bricks and mortar and once in records. Contractors are now required to deliver the LTQR concurrently not years after commissioning and the regulator checks to see that plant changes are incorporated into the facility LTQR. The reason for this change has been increased regulation and the need to understand the construction as an aid to safe decommissioning.

Can you share a specific instance where LTQR played a crucial role in solving a problem or improving safety?

On one occasion we found a radiation shine path from a glovebox unit; due to the location of the unit it was difficult to find the source and as a result the whole building was quarantined. When examining the LTQR pack for the unit in question, it was noticed that during commissioning a weld, whilst within tolerance, was found to be of different quality to others; this was in an area of where it was subject to abrasion due to the process in the glove box. Following examination of the problem glovebox it was confirmed that a weld defect had caused the issue, subsequent checks on other glove boxes revealed potential issues as well, resulting in a repair/upgrade programme.

What's a common mistake companies make when managing LTQR, and how can they avoid it?

Key issues I've found over the years are as follows:

- Plans not completed concurrently with the work and as staff leave projects, information is missing.

- No tracking of LTR information.

- No standard format agreed for LTRs.

- Missing signatures.

- Design drawings not as builts.

What tools or systems do you find most effective for managing LTQR?

A tracker for LTQR information is essential to ensure the LTQR is completed in full. Current electronic collaboration systems make the process of compiling LTQR so much simpler.

How do you ensure that LTQR remain accurate and up to date throughout the project lifecycle?

Planned reviews are essential at key stages and at each point where the asset is altered or updated.

Could you share a success story where LTQR made a significant impact on a project or product?

Hinkley C Marine works – the compilation of the LTQR was entirely electronic and the project used several link digital systems to capture the data concurrently throughout the project. The digital capture meant no signatures and real time updates. Digital capture included a digital twin and enabled the LTQR to be embedded in the digital twin making access for future operators.

What tools, techniques, or strategies do you recommend for maintaining very simple robust traceability in LTQR?

I don't know of any proprietary systems for compiling LTQR but I have worked on the development of bespoke systems using software such as Sharepoint and power BI.

Where do you see the future of LTQR heading in the next five to ten years?

Digital compilation.

2. Interview with Brian Wilson.

Could you introduce yourself and describe your role in the nuclear industry?

Brian Wilson. Quality Manager for Sellafield working on the BEP Project.

What initially attracted you to the field of quality?

Moving into the inspection department was seen as career progression route from the shop floor.

How long have you been working with quality management systems (QMS) and Lifetime Quality Records (LTQR)?

Approximately 30 years.

What are the biggest changes you've witnessed in quality management practices over the years?

Quality management was always part of the manufacturing process. The main change I have witnessed is the need to document the inspection activities and collate the documents and to also document when something goes wrong to enable Quality engineers and Managers to assess workshop processes and modify to ensure repeat problems do not occur.

In your experience, what is the most challenging part of maintaining LTQR? How do you address these challenges on a day-to-day basis?

For me the hardest part of LTR compilation is getting all parties engaged with the importance of the production of documentation and concurrent sign off for these documents. For me, I have had most success with the challenges of LTR compilation by engaging with all members of the workforce and explaining why documentation is needed and how up to date documentation can help in getting product signed off in a timely manner.

How have LTQR evolved during your career, and what changes have you noticed in their importance or complexity?

I have always worked within the nuclear supply chain and documentation has always been present. Some things have changed such as the need for fabricators to be third party approved prior to being allowed to work on nuclear product or paint processes to be approved by Sellafield SMEs before activities are carried out. Generally, this type of upfront approval has increased and as such does impact on the complexity of the LTR that are subsequently produced.

Do you think companies are placing more emphasis on LTQR now than in the past?

Companies are duty bound to place more emphasis on LTR now as they do play a part in them gaining ISO 9001 approval.

Can you share a specific instance where LTQR played a crucial role in solving a problem or improving safety?

I cannot share a specific instance where the LTR has played crucial role in solving a problem but there has been numerous occasions where a Sellafield site based engineer has requested some information or documentation that has been contained within an LTR that I had on record and once supplied has enabled work to continue without delay.

What's a common mistake companies make when managing LTQR, and how can they avoid it?

One of the biggest mistakes I have witnessed is companies putting off the compilation of an LTR under the edict that they will do it later. Inevitably this leads to missing signatures, people not being available anymore to close out an activity and consequently, delay with final sign off and payment for the conclusion of the contract.

What tools or systems do you find most effective for managing LTQR?

I personally have only worked with bespoke systems and Sellafield approved documentation.

How do you ensure that LTQR remain accurate and up to date throughout the project lifecycle?

My own method of managing the accuracy and concurrency of an LTR package is to be pro-active in driving personnel to work to the company's approved procedure for LTRs. This would normally require face-to-face conversations with inspectors to check that visiting QA/QC inspectors have been offered documentation for work carried out on the day and this documentation has been closed out and added to the live LTR prior to their departure.

Could you share a success story where LTQR made a significant impact on a project or product?

I believe that any job where the inspection release certificate has been signed off should be considered as a success. The production of a complete LTR always facilitates the sign off.

Where do you see the future of LTQR heading in the next five to ten years?

I think there will be a move to try to digitalise LTRs going forward which will remove the need for binders and large quantities of paper documents.

What advice would you give to QA teams about maintaining an efficient yet compliant document signing process?

It is paramount that documents are produced and signed off concurrently. My own preferred method was to have any documents pre-drafted prior to the arrival of an inspector which will enable them to close out the activity without delay. Test certificates can be tweaked as and when necessary, but having them available for an inspector to sign off on completion of work witnessed often removes the need to set up a further visit.

If you had to explain LTQR to someone with no background in quality management, how would you describe their importance?

I would describe the importance of a good LTR package as paramount to the success of any work scope.

How does ISO 9001 influence the way you manage Lifetime Quality Records?

Could you provide a specific example of how ISO 9001 has helped streamline or improve LTQR processes in your organisation?

ISO 9001 has been integral within the manufacturing sector for some time now and mandates the need for good record keeping throughout any business. Continuous improvement is a key principle of ISO 9001 and having exemplar LTRs for any product will enable any company to prosper.

3. Interview with Martin Murray

Could you introduce yourself and describe your role in the defence industry?

Martin Murray, Quality Leader at BAE Systems Submarines, I have worked at BAE Systems 39 years. I work mainly on Class 1 nuclear projects doing a mixture of project quality assurance as well as field inspection / surveillance.

What initially attracted you to the field of quality?

Like a lot of people I didn't plan a career in quality, it just kind of happened; however, I wouldn't change it as I love the variety of work and the challenges that it brings.

How long have you been working with quality management systems (QMS) and Lifetime Quality Records (LTQR)?

I joined quality 25 years ago and have been involved in quality management systems throughout that time and first became involved in LTQR a couple of years later when I became resident field engineer for BAES at Rolls-Royce in Derby.

What are the biggest changes you've witnessed in quality management practices over the years?

Inspection techniques and equipment have improved as technology has advanced, the ISO standards have become very generic and watered down during this time. Finally in general the level of skill and experience in industry has dropped.

In your experience, what is the most challenging part of maintaining LTQR?

It's important that the requirements have been properly defined to start with though progressive acceptance is imperative as if anything is wrong it can be rectified earlier in the process and stop potential rework.

How have LTQR evolved during your career, and what changes have you noticed in their importance or complexity?

Introduction of 3.2 material certification has evolved, counterfeit wasn't such a big issue when I started my career but is now. Other improvements include supply chain maps and use of third party inspection bodies.

Do you think companies are placing more emphasis on LTQR now than in the past? Why or why not?

I believe there is a lot more understanding of LTQR these days and a recognition that these play a key role in the assurance and integrity of the product.

Can you share a specific instance where LTQR played a crucial role in solving a problem or improving safety?

Yes, the introduction of supply chain maps and 3.2 material certification improved issues we encountered around steel mills hiding their source of material and also gave us confidence as to the quality of the material and its heritage.

What's a common mistake companies make when managing LTQR, and how can they avoid it?

Trying to compile and assure everything at the end of manufacture rather than progressively can mean they find errors or records missing which can prove difficult to resolve later.

What tools or systems do you find most effective for managing LTQR?

BAE Systems use QASORs (Quality Assurance Statement of Requirements) in submarines for high integrity products; these are basically an index of what records are required and the relevant standards that are applicable. They are very effective and comprehensive, and every record is cross-referenced accordingly.

How do you ensure that LTQR remain accurate and up to date throughout the project lifecycle?

Configuration control and management of new and legacy records is key in ensuring the pack remains up to date at all times.

Could you share a success story where LTQR made a significant impact on a project or product?

The introduction of 3.2 has helped us remain vigilant, and whilst expensive it gives us assurance via the likes of Lloyds Register that the material is sound.

What tools, techniques, or strategies do you recommend for maintaining robust traceability in LTQR?

The use of SAP (Systems, Applications and Products) to record all unique serial numbers, heat, cast and IDs for high integrity products gives us the ability to quickly identify an issue and contain it as opposed to days scanning through records.

Where do you see the future of LTQR heading in the next five to ten years?

The use of digital records and AI systems in controlling and reviewing will no doubt shape the future.

What advice would you give to QA teams about maintaining an efficient yet compliant document signing process?

Get the requirements nailed down at the start and progressively review / accept.

If you had to explain LTQR to someone with no background in quality management, how would you describe their importance?

I have actually done this and use an example of a failure we encountered together with the key LTQR associated to the failure, people do not realise the records associated with high integrity products.

How does ISO 9001 influence the way you manage Lifetime Quality Records?

Could you provide a specific example of how ISO 9001 has helped streamline or improve LTQR processes in your organisation?

In fairness ISO has never really influenced me directly, but it does have some areas where if the requirements for LTQR are challenged it acts as a fall back to suppliers who are accredited to get the best possible outcome.

4. Interview with Craig Mullen

Could you introduce yourself and describe your role in the defence industry?

What initially attracted you to the field of quality?

Craig Mullen, Quality project Engineer at Kinetic Solutions Group (KSG) working on Babcock defence project.

I was initially introduced to quality when I was working in the manufacturing sector producing build and assembly manuals for machinery. Quality was a key part of the assembly process and learning about improvements and problem solving within manufacture I decided to move more into the business management side of the business.

How long have you been working with quality management systems (QMS) and Lifetime Quality Records (LTQR)?

Over 15 years.

What are the biggest changes you've witnessed in quality management practices over the years?

The biggest changes have been the introduction of a Lifetime Quality Record Index. This essentially details all of the deliverables in agreed sections, between the client and supplier. This is particularly helpful when using multiple manufacturers and ensures they all understand how to compile and use the same indexing format.

In your experience, what is the most challenging part of maintaining LTQR?

How do you address these challenges on a day-to-day basis?

The most challenging part is managing the supply chain to ensure that they are concurrently compiling the records as the manufacturing or works are completed. This has always been a huge problem especially if you are coming to the end of the contract and the records haven't been compiled and reviewed. It then becomes a huge effort to try and bring them up to date for handover.

How have LTQR evolved during your career, and what changes have you noticed in their importance or complexity?

LTQR requirements have changed over the years as they are directly related to the requirements from the client. Quality has not always been the main focus; however, in recent years this has been considered to be top of the priority and more focus on the quality records being generated to a high level of requirements.

Do you think companies are placing more emphasis on LTQR now than in the past? Why or why not?

I do believe that companies are placing more emphasis on LTQR for a number of reasons.

- Lack of understanding of the requirements for LTQR previously meant companies ignored the requirement until it was too late.

- Companies were not set up internally to be able to produce LTQR including having the SQEP personnel to produce them.

- Clients wouldn't do the correct supply chain management on suppliers to see if they could provide compliant LTQR.

Can you share a specific instance where LTQR played a crucial role in solving a problem or improving safety?

No specific instances on solving a problem; however, LTQR do improve safety and the records verify the integrity of the SSCs

(SSCs – **S**tructures, **S**ystems, and **C**omponents)

against the design intent. If the SSCs were not tested and records provided to confirm the tests, then the SSCs could fail in operation and lead to issues.

What's a common mistake companies make when managing LTQR, and how can they avoid it?

Not having a structure for the content and not giving the correct level of management/oversight to ensure the records are compiled concurrently.

What tools or systems do you find most effective for managing LTQR?

LTQR Index agreed with the customer and flow down to the suppliers.

How do you ensure that LTQR remain accurate and up to date throughout the project lifecycle?

To ensure that records are accurate and up to date they need to be reviewed concurrently with the project lifecycle by SQEP personnel. The records also need to be signed and dated once reviewed to show that they have been properly examined, validated for accuracy, and approved by qualified individuals at the appropriate stages of the project lifecycle.

Could you share a success story where LTQR made a significant impact on a project or product?

Thirty-plus A4 folders of LTQR was generated for AWE project. These LTQR were two years in the making and on handover from manufacture the project manager commented that the records were market leading in their quality and structure and without this level of completeness the equipment would not have been signed off from manufacture.

What tools, techniques, or strategies do you recommend for maintaining robust traceability in LTQR?

To maintain traceability the records need to be able to be identified to the LTQR and its section. They need to be able to be traced back to the original source, i.e. material.

Where do you see the future of LTQR heading in the next five to ten years?

Electronic generation of LTQR will become more relevant as clients seek a paperless system. The initial records will need to be hard copies, at least in the short term until systems are developed to allow for soft copies of records.

What advice would you give to QA teams about maintaining an efficient yet compliant document signing process?

Wet signing of reviewed records should always be dated to ensure latest records are present. Electronic signatures are becoming more acceptable; however, this format needs to be agreed with the client and is typically a system such as Adobe Acrobat digital signature. By no means should a scanned signature be used to verify a signature.

If you had to explain LTQR to someone with no background in quality management, how would you describe their importance?

LTQR are recorded evidence that the design intent of a product has been met.

How does ISO 9001 influence the way you manage Lifetime Quality Records?

Could you provide a specific example of how ISO 9001 has helped streamline or improve LTQR processes in your organisation?

I am a firm believer of the process approach Plan-Do-Check-Act which is widely used for quality management systems. LTQR are no different to this and applying this method will ensure a more trouble-free generation of LTQR.

APPENDIX C

Checklist for compiling LTQR

This structured format simplifies the process and ensures clarity and consistency when referring to LTQR compilation steps.

Step	Key Actions	Tips
1. Understand the requirements	■ Familiarise with industry regulations. ■ Review customer-specific documentation requirements. ■ Understand internal QMS standards.	Create a requirements summary document for reference.
2. Establish a standard structure	■ Define a clear format: ■ **Cover Page:** Title, project name, reference numbers. ■ **Table of Contents:** Detailed navigation index. ■ **Project Overview:** Scope, objectives, milestones. ■ **Quality Records:** Organise by lifecycle phase: ■ Design Records ■ Procurement Records ■ Manufacturing Records ■ Testing Records	Use templates and consistent numbering systems.

Step	Key Actions	Tips
3. Gather and validate documents	■ Collect records from departments, suppliers, contractors. ■ Validate compliance (e.g. signatures, dates, formats). ■ Cross-check referenced documents are included.	Maintain a checklist of required documents to avoid omissions.
4. Perform Quality Checks	■ Conduct internal reviews for accuracy. ■ Use a checklist to verify completeness. ■ Perform mock audits to identify gaps.	Tailor quality checklists to specific project requirements.
5. Secure storage & accessibility	■ Store digital records in encrypted databases. ■ Maintain physical copies in secure environments, if needed. ■ Implement backup systems to prevent data loss. ■ Track access with logs.	Regularly review permissions to ensure only authorised personnel have access.
6. Communicate with stakeholders	■ Align with internal teams (e.g. project managers, quality teams). ■ Share LTQR with customers, auditors, regulators as required. ■ Collect feedback to improve.	Schedule periodic review meetings to address concerns or suggestions.

APPENDIX D

Terms and definitions

PROCESS

Set of interrelated or interacting activities which transforms inputs into outputs.

PRODUCT

Result of a process. There are four generic product categories, as follows:

- Services (e.g. transport);
- Software (e.g. computer program);
- Hardware (e.g. engine mechanical part);
- Processed materials (e.g lubricant).

PROJECT

Unique process consisting of a set of coordinated and controlled activities with start and finish dates, undertaken to achieve an objective conforming to specific requirements, including the constraints of time, cost, and resources.

QUALITY MANAGEMENT SYSTEM

Management system to direct and control an organisation regarding quality.

QUALITY PLAN

Document specifying which processes, procedures and associated resources will be applied by whom and when, to meet the requirements of a specific project, product, process or contract.

RECORD

Document stating results achieved or providing evidence of activities performed.

SUPPLIER

Organisation or person that provides a product or service.

TECHNICAL QUERY

How subcontractor requests clarification against a technical specification or procedure and may require a Concession to be raised.

CONCESSION

Raised when a product or output is found to be non-conforming with the contract specified requirements. Where appropriate a justification with underpinning evidence will be required to support acceptance.

CE MARKING

Manufacturer's declaration that the product meets the requirements of the applicable EC Directives.

WELDER CERTIFICATION

(Also known as welder qualification) is a process which examines and documents a welder's capability to create welds of acceptable quality following a well defined weld procedure specification.

APPENDIX E

Abbreviations

LTQR – Life Time Quality Records

MTR – Material Traceability Records

WTR – Welding Traceability Records

ITP – Inspection and Test Plan

NDT – Non-Destructive Test

CoC – Certificate of Conformity

WPS – Weld Procedure Specifications

WPQR – Weld Procedure Qualification Record

QCP – Quality Control Plan

BMS – Business Management System

COTS – Commercial Off the Shelf

QA – Quality Assurance

QC – Quality Control

PO – Purchase Order

BOM – Bill Of Materials

ABBREVIATIONS

FAT – Factory Acceptance Test

ICL – Inspection Checklist

IRN – Inspection Release Note

MPI – Magnetic Particle Inspection

O&M – Operation and Maintenance

SAT – Site Acceptance Test

SQEP – Suitably Qualified and Experienced Person

TQ – Technical Query

UT – Ultrasonic Test

www.ingramcontent.com/pod-product-compliance
Lightning Source LLC
Chambersburg PA
CBHW070353200326
41518CB00012B/2223